Department of the Environment

Handling Geographic Information

Report to the Secretary of State for the Environment of the Committee of Enquiry into the Handling of Geographic Information.

Chairman: Lord Chorley

London

Her Majesty's Stationery Office

© *Crown copyright 1987*

First published 1987

Second Impression 1988

ISBN 0 11 752015 2

Front cover photographs courtesy of British Gas plc, Huntings
Technical Services, Laser-Scan Laboratories Ltd, Ministry of
Agriculture, Fisheries and Food, Pinpoint Analysis Ltd.

Contents

List of Appendices

List of Boxes

List of Figures

Foreword

TO THE RT. HON. NICHOLAS RIDLEY MP, SECRETARY OF STATE
FOR THE ENVIRONMENT.

We were appointed by your predecessor, the Rt. Hon. Patrick Jenkin MP on 1
April 1985, with the following terms of reference:

> 'To advise the Secretary of State for the Environment within two years on
> the future handling of geographic information in the United Kingdom,
> taking account of modern developments in information technology and of
> market need.'

We now have pleasure in submitting our report which is unanimous.

Evidence

We issued an open invitation to submit evidence in June 1985 and sent this to
public and private sector organisations with an interest in the subject. We also
wrote to a number of relevant organisations and individuals in other countries
and to professional and technical journals in the United Kingdom. A copy of
the invitation to submit evidence is at Appendix 1.

We received written evidence from nearly 400 organisations and individuals
and took oral evidence from 26 organisations; these are listed in Appendix 2.
Copies of this evidence will be deposited in the Department of the Environment
Library.

We also asked individuals to give presentations to the Committee on specific
topics and we are grateful to Dr Peter Dale, Professor J.T. Coppock, Mr
Michael Brand, Mr Chris Denham and Professor Howard Newby for their help
in this way.

Studies and reviews

To assist us in our work the Department of the Environment commissioned, at
our request, 4 small studies or reviews: Logica Space and Defence Systems Ltd
prepared a review of developments in Information Technology relevant to
Geographic Information Systems; Dr Stan Openshaw of Newcastle University
prepared a review of spatial units and locational referencing; Ecotec Ltd carried
out a study of market demand for geographic information in the private sector;
Tomlinson Associates presented a review of North American experience of
Geographic Information Systems. Parts of the reviews by Tomlinson
Associates and Dr Openshaw are included as Appendices 6 and 7 of our
report.

Meetings and visits

We held 22 meetings. Members of the Committee also made visits to relevant
organisations and these are listed in Appendix 2.

International contacts

In addition to receiving written comments from organisations and individuals
from abroad we took advantage of the presence in London of overseas
specialists, attending the Auto Carto London conference in September 1986, to
organise a small international seminar. A group of Committee members also

visited the University of Utrecht to talk with Dutch experts about the organisational arrangements for Geographic Information Systems in Holland. The Chairman attended a conference of the American Congress on Surveying and Mapping and the Society for Photogrammetry & Remote Sensing in Washington in March 1986 and whilst in North America had discussions with officials at the US Geological Survey, the US National Science Foundation and the Canadian Land Data System.

Acknowledgements   We are most grateful to all the organisations and individuals who either submitted evidence to us or gave of their time and expertise to help us with our task. This was invaluable.

We were much helped in our work by the wise counsel of the DOE Assessors to the Committee, initially Dr Michael Richardson, until his retirement in November 1985 (when he joined the Committee), and subsequently Mr David Wroe. Government Departments were kept in touch with our work and provided useful feedback through a working group chaired by the DOE Assessor; we are indebted to them for the helpful suggestions they made.

Finally, we warmly thank our Secretariat: John Metcalf (Secretary), Christabel Myers, and John Troughton (Technical Adviser). They had to digest and analyse an enormous volume and range of evidence in a difficult technical field. Their help was invaluable. Our thanks also go to Joyce Fallconi, David Lewis and Jennifer Schwitter who provided essential support.

Roger Chorley (Chairman)	Walter Smith (Deputy Chairman)
Jean Balfour	Ian Gilfoyle
David Rhind	Michael Richardson
Frank Russell	Russel Schiller
Gurmukh Singh	David Sinker
John Townshend	

Summary of conclusions

General

1   'Geographic information' is information which can be related to specific locations on the Earth. It covers an enormous range, including the distribution of natural resources, descriptions of infrastructure, patterns of land use and the health, wealth, employment, housing and voting habits of people.

2   Large sums of money are spent by Government, commerce and industry, the utilities, the armed forces, and others in collecting and using it. Much human activity depends upon the effective handling of such information − its collection, processing, storage and retrieval − but it is the ability of computer systems (Geographic Information Systems or GIS) to integrate these functions and to deal with the locational character of geographic information that makes them such potentially powerful tools.

3   Falling computer costs and the rapid development of new systems now offer a real possibility for improvement in the use and handling of geographic information, in the interests of better management, use of resources, planning and decision-making.

4   Our Enquiry has been concerned with identifying the means whereby these benefits may be secured and any barriers or constraints removed. The principal barriers reflect:

● the problems of dealing with a great diversity of users and uses

● the fact that Geographic Information Systems are only tools, albeit essential ones, for better decision-making. Any lack of awareness of potential benefits can therefore easily result in failure to recognise the central importance of these systems.

Costs

5   The conversion of existing paper records into digital form is a significant cost. It is not easily automated and cheaper conversion processes are required. For new data, direct recording in digital form eliminates this problem.

6   Manipulating geographic information is costly because it requires considerable computer processing power. But computer processing and related costs will continue to fall dramatically and systems will become more widely available.

Availability of data

7   Availability of data in digital form is a prerequisite for using computer handling techniques.

8   Rapid conversion of Ordnance Survey (OS) basic scales maps to digital form is a priority; the present OS programme is too slow to meet user needs.

9   If a more rapid OS map conversion programme is not started soon, there is a serious risk of a costly duplication of effort and proliferating standards.

10   Automated map digitising techniques offer the possibility of a more rapid solution. OS needs to test these techniques in collaboration with the main users, particularly the utilities.

11   Government Departments hold much property and socio-economic data which are wanted by outside users. Often such data are not made available for confidentiality or cost reasons. The confidentiality restrictions on some Land Registry and Valuation Office data are unnecessary. Emphasis on gross cost limits rather than net costs unduly restricts the marketing of data by Government Departments.

12   Government Departments and other organisations should adopt a positive approach to the marketing of their data. To maintain flexibility, data should be held in as disaggregated a form as possible. Data registers are needed to provide potential users with information about information.

Linking data   13   The benefits of a Geographic Information System depend on linking different data sets together. Apart from general Information Technology standards linkage requires two things.

14   First, locational references (we recommend the use of National Grid coordinates) are required, together with standard 'spatial' units for holding and releasing data (we recommend the use of unit post codes for most purposes). The 1991 Census of Population will have a crucial role to play in securing the widespread use of standard spatial units. It is essential that the plans for the Census make provision for postcode referencing.

15   Second, data documentation and exchange standards are required. For map-based data a National Working Party under OS leadership has produced draft standards and comparable initiatives are required for other data.

Awareness and human and organisational problems   16   Many users and potential users are not sufficiently aware of the benefits of Geographic Information Systems. A major promotional exercise is required to maximise returns on the existing national investment in geographic information.

17   There is also a need to increase substantially, and at all levels, the provision of trained personnel.

18   The human and organisational problems of achieving change are also an important barrier to development. Information needs to be seen as a corporate resource and be more widely shared between departments and organisations.

Research and development   19   There are a number of research and development (R&D) tasks which are specific to Geographic Information Systems although benefits are also derived from much of the research and development undertaken in the general field of Information Technology.

20   At present, insufficient funds are allocated for these specific research and development tasks. The reasons for this include the lack of any machinery to bring the wide diversity of users together, the subject falling between the interests of the different Research Councils, and the lack of awareness of the subject of those allocating R&D funds.

Coordination   21   The full potential of Geographic Information Systems will only be secured by coordinating the interests of the widely dispersed community of users. At present, no single body has a responsibility to secure this coordination.

22   We therefore recommend setting up a body, a Centre for Geographic Information, which would have three main tasks:

- to provide a focus and forum for user groups
- to carry out promotional activities
- to oversee progress and to submit proposals for developing national policy.

23   The Government, as the main supplier and user of geographic information, will have a central role to play. We therefore emphasise the need for the Government to give a lead.

Immediate tasks   24   We highlight four key tasks which need to be put in hand now if important opportunities are not to be lost:

1   Rapid conversion of the OS basic scales map series to digital form in collaboration with the main users.

2   More widespread use of post codes as standard units for holding and releasing socio-economic data. The plans for the 1991 Population Census are crucial in this respect.

3   More widespread use of the National Grid referencing systems to link different data sets.

4   Setting up a Centre for Geographic Information to carry matters forward in promotion, training, and bringing users together and to submit proposals for developing national policy.

SECTION I

Introduction

1 Geographic information and the computer

1.1 'Geographic information' is information which can be related to specific locations on the Earth. It covers an enormous range, including the distribution of natural resources, the incidence of pollutants, descriptions of infrastructure such as buildings, utility and transport services, patterns of land use and the health, wealth, employment, housing and voting habits of people.

1.2 Most human activity depends on geographic information: on knowing where things are and understanding how they relate to each other. Many aspects of decision making − for management, planning and investment − by government and the commercial world depend on it. For example, government uses geographic information on the location of particular groups of people − such as the unemployed and elderly − to direct services, businesses use it to identify likely customers and to site their depots and branches, and it is essential to the operations of the armed forces.

1.3 Large sums of money are spent on collecting geographic information, but using it to its full potential is often difficult. For example, linking together separately collected pieces of information relating to the same location is sometimes impossible because of the different ways in which that location is described.

1.4 In addition, the large amount of information required even for simple problems − such as estimating the number of UK households within 10 kilometres of a stretch of motorway − often makes manual analysis tedious and error-prone. More complex problems − such as predicting the effect on land-use of changes in the Common Agricultural Policy − might require the modelling of alternatives. This sort of operation, which is often an essential part of the decision making process, requires the use of sophisticated computer systems.

The impact of new technology

1.5 Until recently, the computer systems to handle such information − 'Geographic Information Systems' (GIS) − were too expensive and cumbersome for most users. Two developments are rapidly changing this. The first is the dramatic fall in computer costs: processing costs have fallen by a factor of one hundred in the last ten years, with the expectation of similar changes in the future. Today, even a desk-top micro-computer can hold useful amounts of data and carry out many of the manipulations that are required. In addition, the growth in packaged software has significantly reduced system development costs. Along with this has been the increasing ease of using computer systems.

1.6 The second and complementary development, which stems in part from the fall in computer costs, is the increased availability of commonly used data in digital form, such as that from the Census of Population; for each user to have to convert such data from paper to digital records is slow and expensive.

1.7 These developments do not resolve all the problems of handling geographic information. Nevertheless, they have aided the development of an extremely important tool — the Geographic Information System — which has considerable potential for development. Such a system is as significant to spatial analysis as the inventions of the microscope and telescope were to science, the computer to economics, and the printing press to information dissemination. It is the biggest step forward in the handling of geographic information since the invention of the map.

The need for a review of geographic information handling

1.8 The opportunities afforded by the new technology for handling geographic information and the problems involved in realising them mirror in many ways those brought about by developments in Information Technology in general. Many studies of the impact of these developments have already been carried out by Government and others.

1.9 However, geographic information and the technology for handling it have a number of distinctive features which merit a separate study, including:

1 the 'spatial' dimension of the information i.e. its location. Standard ways of describing the location of geographic information or 'spatial data', and special software are required to permit the linking of such data sets and the examination of relationships between them.

2 the use of geographic information in the form of maps, which are an essential display tool for many purposes. With the new technology maps can be purpose designed for specific applications and are thus considerably more effective. But the design and display of maps in digital form make significant demands on software.

3 the need to convert from paper to digital form a vast backlog of existing maps, both topographic and thematic, which are the sole source of important spatial data. This conversion process requires the development of special technology.

4 the wide range of geographic information and its uses encompass a vast number of users and other interested parties such as data suppliers. This diversity causes difficulties in communication between them and hence delays the development and application of new technology.

1.10 Concern that — because of such distinctive features — geographic information might not be fully exploited using the new technology, led the House of Lords Select Committee on Science and Technology to recommend in 1983, the establishment of our Enquiry. The Government accepted that recommendation (Refs. 1 and 2). In the remainder of this Chapter, we outline the scope of our Enquiry, the benefits which we believe will arise from the application of geographic information technology, and key factors critical to realising these benefits.

The scope of our Enquiry

1.11 Our Enquiry covered the United Kingdom. We found that the countries within it generally face similar problems and opportunities arising from the use of geographic information: in our report, we highlight the few significant differences between them. The possible range of our Enquiry was vast. In the interests of completing our task, we deliberately limited its scope and focussed on land based matters. We have given no detailed consideration to the marine and atmospheric environments.

1.12 We have accordingly concentrated on the uses of the following types of geographic information:

1 land and property data. Here the main data sets are those on property titles maintained by the Land Registries and local authorities, rateable values and property transactions held by the Valuation Office, and planning applications and consents held on local authority planning registers.

2 rural resources and environmentally related data. This broad area includes land use, natural resources, ecological and environmental data. An increasingly important source of such data is satellite or remotely sensed imagery.

3 infrastructure data. This includes the layout and attributes of a utility's pipes and cables and of transport networks.

4 socio-economic data. In this area perhaps the most generally used data set is the decennial Census of Population. Other important data sets describe employment, unemployment, health and expenditure.

1.13 All these data are 'spatial data' since they can be referenced to a geographic location. For Great Britain, a referencing framework is provided by the National Grid system to which Ordnance Survey (OS) mapping data is referenced. For Northern Ireland, the Irish Grid provides an equivalent framework.

1.14 We have examined past and current uses of such data and ways of handling them so as to provide a basis for assessing future uses. In the last 15 years a number of Geographic Information Systems have been developed. These ventures have not been without difficulty, stemming from over-ambition, insufficient attention to user needs, conservatism of users, and over-optimistic estimates of data conversion and system development costs. From a national point-of-view, these endeavours provide essential experience in learning to use new technologies. The important point now is to draw the lessons from this experience.

1.15 It is also important to learn from overseas experience. Undoubtedly the most impressive developments are to be seen in North America which has been the power-house in the creation of computer hardware and software.

The benefits of Geographic Information Systems

1.16 Well-designed Geographic Information Systems now have the capability of providing:

1 quick and easy access to large volumes of data;

2 the ability to select detail by area or theme; to link or merge one data set with another; to analyse spatial characteristics of data; to search for particular characteristics or features in an area; to update data quickly and cheaply; to model data and assess alternatives; and

3 new and flexible forms of output — such as maps, graphs, address lists and summary statistics — tailored to meet particular needs.

1.17 For public sector users, such systems provide a means of more efficient provision of services; to private sector users, they offer a valuable way of increasing competitiveness — indeed, their adoption may become essential to maintaining competitiveness. Their benefits will be most apparent to organisations which already use and rely on geographic information. But it is likely that their use will quickly spread to other organisations. There is already evidence of this happening in the private sector — for example, in the analysis of customers and markets.

1.18   Geographic Information Systems and digital geographic information also present a valuable commerical opportunity for the provision of new computer based products and value added information services. The United Kingdom is quite advanced in certain areas of geographic information technology and its applications; that lead could be used to compete successfully in world markets.

1.19   At present, Geographic Information Systems may appear to be of modest benefit to some applications. This is in part because of the still high costs of computer systems including purchase of systems and conversion of paper records, and partly due to the difficulty of quantifying the benefits of better information. In addition, many organisations seem not to value information sufficiently as a vital corporate resource whose use should be maximised to secure the greatest return on the investment made in collecting it. In the longer term as costs fall and the benefits are better understood, these systems will more clearly result in cost effective operations, both for planning and managing resources.

The future   1.20   We are convinced that a fundamental change towards the routine use of geographic data on computers has already begun. The consequences of this cannot be predicted except in general terms and by analogy. The introduction of computer accounting systems 25 years ago was hesitant at first and many mistakes were made. However, costs fell, the value of new forms of information from the systems was discovered and, in due course, these systems became commonplace in quite small organisations. A return to manual systems is now inconceivable. We believe the use of Geographic Information Systems will develop in a similar fashion.

1.21   It is not easy to predict the rate of growth in the use of Geographic Information Systems. Some reputable estimates suggest an annual US market of $3 billion a year by 1990, several times larger than the present figure. We suspect growth in this country, except in the utilities and a few other specific areas, may be rather slower unless we become more receptive to change and appreciative of the benefits of better information.

1.22   Even while we have been sitting, the technology has moved forward and all the predictions strongly suggest continuing rapid development of hardware, software and telecommunications. We believe this to be a necessary, though not sufficient, condition for the take-up of Geographic Information Systems to increase rapidly.

1.23   A factor which is critical in determining the take-up of new technologies is that of user awareness. This is particularly the case when quite new applications may be facilitated by the technology — a factor which is important in all sectors of Information Technology and has been recognised in many other reports.

1.24   A further critical factor is the availability of digital spatial data in a form suitable for use in particular applications and to facilitate the linking of data sets. An underlying issue here is that the main data collectors and holders may not necessarily wish to exploit the spatial dimension of their data themselves and may not therefore provide the data in a form which will facilitate spatial analysis by others.

1.25   Government will have an essential part to play in realising the benefits brought by the wider use of Geographic Information Systems. It is the biggest user and the largest single supplier of geographic information in the United Kingdom.

1.26   Some Government data — such as that collected in the Census of Population — are of perceived direct value to the commercial sector or to local government. Less well recognised is the value of rural resources and land use data which can be applied at national, regional and local levels. A Government lead — through using and encouraging the use of this information within Geographic Information Systems — is crucial, for example, to planning for rural communities and the leisure and recreation of our largely urban population.

1.27   The very diversity of users and applications causes difficulties in making recommendations to speed take-up of the technology. Nonetheless, our main task has been to suggest measures which will bring common benefit by removing the barriers to change. Some of these barriers are human and institutional, some are administrative and legal, some are financial, and others are of a technical nature.

1.28   Our proposed measures will require some investment; the majority of this will come from commerce and users themselves, with a relatively modest contribution from Government. This investment should give rise to major opportunities but these can only be realised under conditions where innovation is commonplace and welcomed.

The structure of our report

1.29   Our report is organised on the following lines:

Section 2:

We examine the current uses of geographic information and Geographic Information Systems, and identify the major factors likely to influence take-up.

Section 3:

We then consider in more detail the barriers to development and the actions needed to overcome them. We deal in turn with:

1   the provision of digital topographic mapping by OS, including the crucial issue of how to achieve rapid national coverage and related copyright and charging issues (Chapter 4);

2   the availability of Government data, including the need to make more data available and in a form which meets users' needs; the difficulty of finding out what data are available; and inhibiting factors such as confidentiality, administrative attitudes, and costs (Chaper 5);

3   how different data sets can be linked; issues of central importance here being the need for locational referencing, for standard spatial units by which data are held and provided, and for data transfer standards to facilitate the exchange of geographic information (Chapter 6);

4   what needs to be done to encourage awareness of geographic information technology among users (Chapter 7) and to increase the supply of skilled personnel (Chapter 8); and

5   the nature of the research and development tasks which need to be tackled and the respective roles of systems suppliers, Government, the Research Councils and academia (Chapter 9).

Finally, we consider the role of Government and the organisational machinery that is needed to ensure the most effective use of modern data handling techniques. We conclude that there is a need for a central body, a Centre for Geographic Information. This would provide a focus for the various different interest groups, be responsible for promoting the use of the new techniques and have a general remit to monitor progress and help develop national policies (Chapter 10).

1.30   We received a large amount of evidence in the course of our Enquiry, most of it of high quality. In order to make at least some of it more accessible to a wider audience, we have included selected extracts as Appendix 10 of this report.

SECTION II

Current and future use of geographic information

2 The use of geographic information today

2.1 To provide a basis for our assessment of requirements for geographic information and Geographic Information Systems, we assembled a considerable body of information on their existing use and on the potential for growth. We outline below the major UK sources of data and the data handling equipment and software available, who uses them, and for what purposes. We further illustrate current practice through a brief review of overseas experience and a number of case studies.

Data 2.2 The majority of spatial data used by organisations are generated internally. For example, most locationally referenced data held and used by the utilities are their own plant and mains records. However, there are a number of data sets providing national coverage which many organisations require for use on their own or, more often, to link to internally generated data.

2.3 By far the largest suppliers of national spatial data sets are Government Departments − such as the Ordnance Survey (OS), the Office of Population Censuses and Surveys, the Department of Employment (D.Emp), and the Agricultural Departments − and other Government funded organisations such as the Research Councils, Health Authorities and Tourist Boards. An indication of the types of data which are most widely available and their sources is given in *Box 2A;* a more detailed list of spatial data held by Government Departments and local authorities − based on submissions of evidence − is included in Appendix 3.

2.4 An increasing number of major data sets is being made available in digital form as a by-product of suppliers computerising their data holdings to meet their own needs. Most Government Departments have or are planning to computerise their major data holdings. Several are in the form of integrated databases, such as those being set up by the Military Survey Directorate of the Ministry of Defence (MOD), the Ordnance Survey Northern Ireland, the Department of Transport (DTp) and the D.Emp.

2.5 Aerial photographs or imagery taken from aircraft continue to be an important form of spatial data used by many organisations. An increasingly significant source of imagery is the satellite. The manner in which satellite or remotely sensed data are collected and some current applications are described in *Box 2B*.

2.6 Remotely sensed data are primarily of value as a source of environmental information. The resolution of such data from new satellites is higher than before *(Fig. 2.1)* but it is currently sufficient for mapping only at small scales. Moreover, for many UK applications such as crop monitoring, the frequency of data collection remains insufficient due to the high incidence of cloud cover. This is likely to remain the case until the advent of satellite-borne radars such as that on the ERS-1 satellite. Remotely sensed data for overseas countries are often the best quality information available and are of considerable importance

Box 2A: Major types and sources of spatial data

Land and property

Large Scale Maps	Ordnance Survey (*Northern Ireland)
Ownership	HM Land Registry (*Scotland, Northern Ireland)
Value	Inland Revenue/Valuation Office (*Northern Ireland)
Land Use (including land use change, planning permissions, and availability)	Department of Environment (*Scotland, Northern Ireland), Local Authorities
Housing Stock, Condition, Building Rates	Department of Environment (*Scotland, Northern Ireland), Local Authorities

Socio-economic

Population Characteristics	Office of Population Censuses & Surveys (*Scotland, Northern Ireland), Local Authorities
Medical Statistics	Office of Population Censuses & Surveys (*Scotland, Northern Ireland), Department of Health & Social Security (*Northern Ireland), National Health Service, Scottish Health Service
Labour Market (including employment, unemployment, vacancies)	Department of Employment (*Northern Ireland), Manpower Services Commission
Business Statistics	Department of Trade & Industry/Business Statistics Office (*Scotland, Northern Ireland)
Social Benefits	Department of Health & Social Security (*Northern Ireland)
Crime Statistics	Home Office (*Northern Ireland), Royal Ulster Constabulary, Local Authorities
Tourism and Leisure	National and Regional Tourist Boards, Countryside Commissions, National Park Authorities, Local Authorities

Land use, rural resources and environmental

Small Scale Maps	Ordnance Survey (*Northern Ireland)
Agriculture	Ministry of Agriculture, Fisheries & Food (*Wales, Scotland, Northern Ireland)

Natural Resources	Natural Environment Research Council/British Geological Survey/Institute for Terrestrial Ecology, Soil Survey for England & Wales, Macaulay Institute, Survey Companies e.g. BKS Surveys UK, Oil Companies, National Coal Board
Forestry	Forestry Commission (*Northern Ireland)
Environmental Quality	Department of Trade & Industry/Warren Spring Laboratory, Department of the Environment (*Northern Ireland), National Remote Sensing Centre, Local Authorities
Atmospheric Conditions	Ministry of Defence/Meteorological Office, National Remote Sensing Centre

Infrastructure data

Utilities' Networks	Electricity Boards, Water Authorities, British Gas, British Telecom, Department of the Environment Northern Ireland
TV Networks	TV companies
Road Networks	Department of Transport (*Scotland, Wales, Northern Ireland), Local Authorities
Other Transport Networks	British Rail, Northern Ireland Railways, London Regional Transport, Water Authorities

Sea and air

Hydrographic & Coastal Topography	Ministry of Defence/Hydrographer, Ordnance Survey (*Northern Ireland), Port Authorities
Marine and Oceanographic	Natural Environment Research Council/Marine Information & Advisory Service, Ministry of Defence/Hydrographer, Ministry of Agriculture, Fisheries & Food (*Scotland, Northern Ireland), Oil Companies, Geological Survey Companies
Air Navigation Maps	Ministry of Defence/Military Survey, Civil Aviation Authority

*Equivalent source of data from Government offices in Wales, Scotland, Northern Ireland as indicated.

Note: More details of these data sets are given in Appendix 3.

Box 2B: **Remotely sensed data and applications**

1   The use of remotely sensed data from both airborne and satellite sensors is rapidly increasing, as are the applications of such data by earth scientists, engineers and planners. The USA Landsat and the French SPOT remote sensing satellites generate high resolution space imagery; lower resolution imagery is provided by meteorological satellites. Imaging radar and other microwave remote sensing instruments are currently under development by the European Space Agency.

2   The Landsat Thematic Mapper instrument operates in seven spectral bands producing imagery built up from individual picture elements (pixels) of 30m x 30m on the ground to form scenes of area 185km x 185km and having a repeat cycle of observation of 16 days. The SPOT instrument operating in three spectral bands produces 20m x 20m pixels to form scenes of 60km x 60km. This instrument also has the capability to produce panchromatic images with 10m x 10m pixels and can observe the ground by oblique viewing to increase the repeat cycle to 5 days. The oblique viewing also provides stereoscopic observations of topographic relief.

3   Applications of satellite and airborne imagery relate knowledge of the Earth's surface and land cover which can be deduced from the reflected solar radiation and observations of thermal infra-red radiation. Cloud cover data turn-around time and difficulties of interpretation act as constraints. Some typical land and coastal area applications are:

Geology/Geophysics/Geomorphology: Remote sensing imagery provides a useful data source for geological structure and tectonics and, at a local level, information for discriminating between different minerals. Imagery is also used for mapping structure at global, regional and local scales as an aid to mineral and hydrocarbon exploration.

Land Use Planning/Agriculture/Ecology: Applications in this area are concerned with soils, crops, forests, natural vegetation and other land cover features. They include soil survey, crop mapping, forest inventory, habitat mapping, vegetation index observations on regional scales and land cover mapping for landscape change detection.

Hydrology/Hydrography: Applications here include rainfall monitoring, distribution of snow and ice, soil moisture levels and variations, water quality and suspended sediment loads in coastal and inland waters and coastal bathymetry.

Cartography: Remotely sensed imagery, including space borne photography, has contributed to the mapping of overseas areas at scales of 1:50,000 and smaller. Map revision can also be based on the use of space imagery to detect feature changes which assist local ground surveys.

4   Remotely sensed data are rarely used as sole sources of information in the applications listed above. Aerial photography, topographic maps and ground survey results are all used in combination with remotely sensed data to assist with the analysis and interpretation of the features of interest. Image processing systems provide a means for handling such information and for combining raster-based remotely sensed data with other spatially referenced data such as digital terrain models and vector-based digitised maps.

Fig. 2.1 **Improved resolution of satellite data**

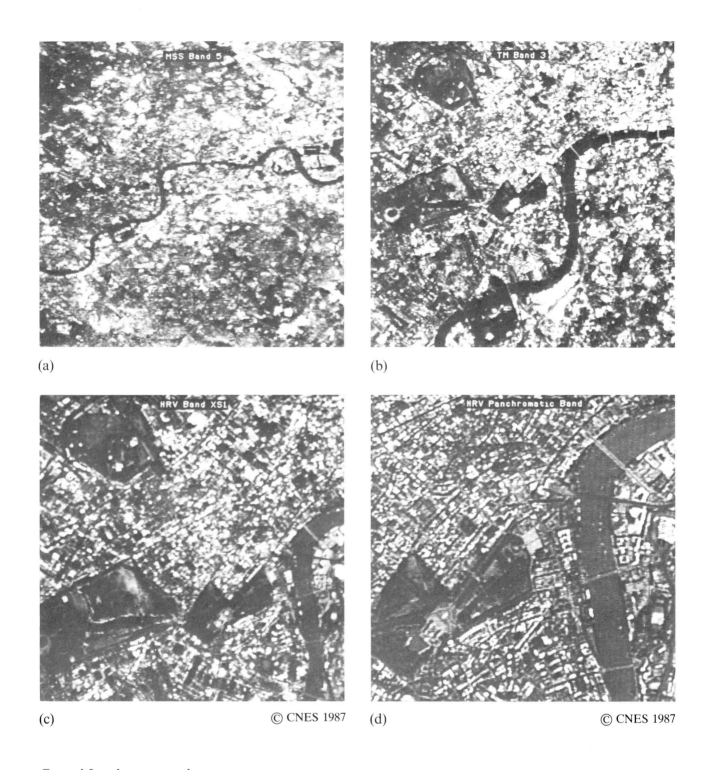

(a)

(b)

(c)   © CNES 1987

(d)   © CNES 1987

Central London as seen by
(a) the Landsat Multispectral Scanner
(b) the Landsat Thematic Mapper
(c) the SPOT HRV in multispectral mode
(d) the SPOT HRV in panchromatic mode.

to UK companies undertaking work in areas such as mineral exploration and foreign aid programmes.

2.7   The value of remotely sensed data is usually realised through their integration with data collected by more conventional means. However, it is currently difficult to input and integrate the two sorts of data (raster and vector) in Geographic Information Systems. Measures to advance the exploitation and development of remote sensing techniques are increasingly being organised through the recently established British National Space Centre. One project it is currently funding is a feasibility study into a Geographic Information System having the capacity to handle raster and vector data.

Computer handling techniques

2.8   We outlined in Chapter I how the use of geographic information could be transformed by employing computer handling techniques. To achieve this, computer equipment and software are required to 'capture' the data, including the conversion of paper maps and text into digital form; store and manipulate the data held in the computer; and retrieve and display the date — for example — via a visual display unit *(Fig. 2.2)*. Communications networks are needed to permit the linking of one computer with others so as to transfer data. These techniques and the types of equipment and software required are described in more detail in Appendix 4.

2.9   Until recently, Geographic Information Systems incorporating some or all of these techniques were purpose-designed and generally developed in-house. There are now several 'off-the-shelf' systems available which vary in the emphasis placed on particular techniques — such as powerful graphics or analytical capabilities, or the ability to integrate map and image data. Most systems are American although UK firms have been successful in meeting the requirements of some users.

2.10   Although the cost of full systems is falling rapidly, it can still be expensive to install a new one, particularly when it involves any rearrangement of data held in a form — paper or computer — which is incompatible with the new system. It is possible to use general purpose computer equipment plus 'add-ons' such as special software and plotters to handle spatial data, provided the data are held in a suitable form on the computer. Geographic Information Systems for personal computers are being developed, but are currently limited in data storage capacity.

Users and uses

2.11   As an aid to decision taking and resource management, geographic information is used by public and private sector organisations in a wide range of applications. The present employment of Geographic Information Systems in full or part varies both by type of user and application area. Requirements for spatial data and computer handling techniques in the main applications areas are outlined below.

Planning and managing public services

2.12   Both central and local government use geographic information to assist in planning and managing the services they provide. Examples of how geographic information technology is currently being used to handle Government data are given in Case Study Example I, at the end of this chapter, and to handle local authority data in Example 2.

2.13   Central Government uses a wide range of geographic information. For example, Census of Population and Employment data are used to help establish priorities for inner city initiatives, local government rate support grants and

Fig. 2.2 **Elements of an information system**

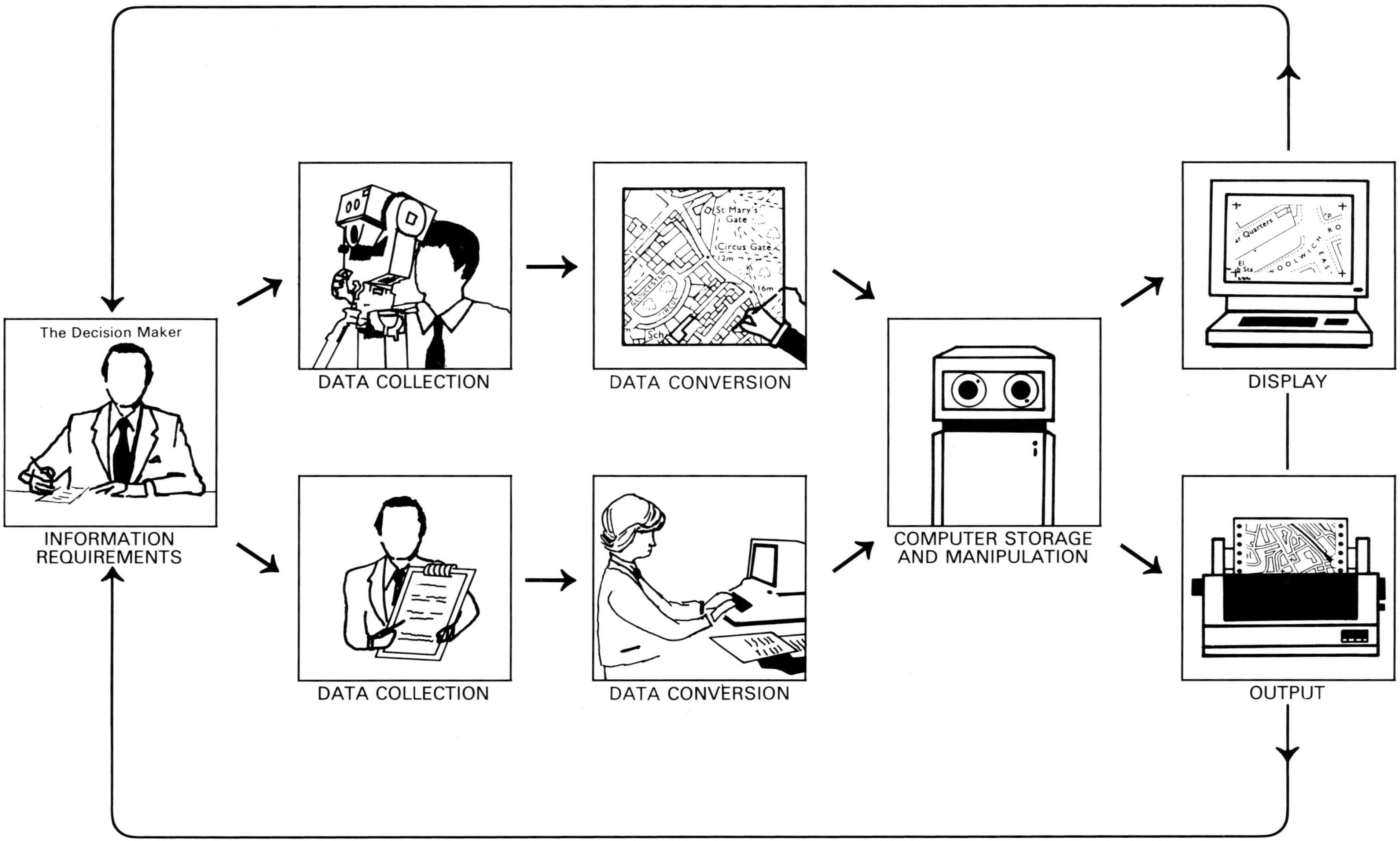

assisted areas policies. Some Geographic Information Systems have already been developed mainly to handle transport, socio-economic, and land and property data. A number of these are predominantly for internal management purposes: for example, the planning appeals information system used within the Department of the Environment (DOE). Others are of more national significance: for example, road transport information systems developed by the DTp and the Scottish Office (see para 2.31) and the Manpower Services Commission's National On-line Manpower Information System (NOMIS), which provides easy and rapid access to data on labour markets for Government and other users.

2.14   Local authorities require information on numerous aspects of the local areas for which they provide services. *Box 2C* outlines the types of data local authorities require and the uses that might be made of it. Although local authorities hold much of this information on computer, the use of Geographic Information Systems to handle it is relatively limited. They have been little applied in the education or social services areas or as a corporate tool to help management of resources across departments.

2.15   The area where spatial techniques have been most applied in local government is in land and property information systems. Since the 1970s, this area has provided a focus for various Geographic Information Systems. A number of local authorities now operate systems which use the address as a basis for holding their data and can perform spatial searching.

2.16   More recently, however, consideration has been given by the parties concerned to linking local authorities data on local land and property with Government data on land ownership. The second report of the Conveyancing Committee (the Farrand Committee), published in 1985 (Ref. 3), recognised that, as computerisation of local authority and Land Registry records proceeded, there would be advantages in establishing computer to computer links to permit access from any single point to all other registers and records throughout the country. We support this proposition. In the short term, it would be of value to those interested in land ownership and development questions including both central and local government, other public bodies such as the Boundary Commissions, and the private sector. In the longer term, it might provide a framework for integrating some other land and property related data such as non-confidential data held by the Valuation Office (VO). Appendix 5 provides more information on land and property data bases and computerised systems.

Defence and security systems

2.17   An important use of geographic information and new technology for handling it is in command and control systems for defence and civil security, including police and fire operations. In these areas, users are often required to assimilate large amounts of information from different sources, in order to make instant decisions and communicate them to others. In a military context, the data required might include the location of military units, supply stations and airfields for army manoeuvres, and terrain height and atmospheric conditions for flight navigation or missile guidance. For civilian users, data used might include the locations of police stations, fire hydrants, properties containing toxic materials, generators, water supplies and the street network.

2.18   Sophisticated computer systems are being developed to help the command and control process, particularly in the defence area. Paper maps, photographs and alphanumeric data stored in computer form can be used to direct resources with greater accuracy, provide a simulated physical overview

Box 2C: **Local Government uses of geographic information**

Monitoring changes in resources (land and building, equipment and infrastructure) and conditions (economic, social, demographic, environmental)

Forecasting changes in housing requirements, in school rolls, in travel patterns, in the economy and in the demand for land, leisure, and community services.

Service planning through identifying and forecasting changes in patterns of need for services and investments as a basis for the delivery of services and deployment of resources. This will determine both the scale of provision and its location; it will also highlight areas of social deprivation.

Resource management e.g. building maintenance, refuse collection, grass cutting, route scheduling of supplies vehicles, mobile libraries, social service ambulances.

Transport network management including provision and maintenance of highways, public transport schedules, school transport, street cleaning.

Public protection and security systems e.g. police command and control systems, definition of police beats, location of fire hydrants, patterns of crime and incidents of fire.

Property development and investment including the preparation of development plans; assessing land potential and preparing property registers; promoting industrial development; rural resource management.

Education use of a wide range of data for teaching purposes, including the use of demonstration software packages.

23

such as in terrain modelling *(Fig. 2.3)*, and overlay locations of important resources. Data compression techniques have been developed by the MOD to enable storage and use of data on portable machinery used, for example, in aircraft. These developments may have valuable applications in civilian areas.

Land use and rural resources management

2.19 The use of rural land for agriculture, forestry, conservation and recreation and the changes between these activities or between them and other forms of land use is primarily the concern of Government Departments — the Agricultural Departments, the DOE and the Scottish and Welsh Offices. The management of rural land, however, lies predominantly in small private ownerships, though agencies such as the Forestry Commisssion and the Nature Conservancy Council both own and manage land and carry out regulatory and monitoring functions.

2.20 The Agricultural Censuses in Britain, collected by the Ministry of Agriculture, Fisheries & Food (MAFF) and the Department of Agriculture & Fisheries for Scotland (DAFS), and information about landscape and land use change, depend on frequent data collection which is costly. Land managers and others with land based interests require OS 1:10,000 scale maps which are not currently in digital form. A further source of data is the Soil Survey of England & Wales which has established, with the Agricultural Development Advisory Service of MAFF, a computerised Land Information System (LANDIS). In Scotland the Macaulay Institute for Soil Research provides computerised data for soils and land capability for agriculture which is used by DAFS. Because of increasing interest in recording land use change and the high cost of data collection and updating, sampling techniques and the use of satellite data are being investigated. Examples of the latter are to be found in Strathclyde and Grampian Regional Councils and in the DOE.

2.21 There have been some feasibility studies for creating integrated rural resources and land use information systems — including the Rural Land Use Information System (RLUIS), which was set up by the Scottish Development Office and others in the 1970's but not developed. More recently, the Rural Wales Terrestrial Database (WALTER) feasibility study (Ref. 4) was carried out for the Welsh Office and others who have agreed to initiate a second phase of the study in conjunction with the Economic & Social Research Council Regional Research Laboratory at the University of Wales Institute of Science & Technology.

2.22 So far, relatively little use has been made of comprehensive Geographic Information Systems to manage rural resources. This is partly due to attitudes within Government Departments, the pattern of small land ownerships and the costs of the systems. However, spatial data handling techniques are being introduced by a limited number of organisations such as the Forestry Commission to help manage its forest enterprise and carry out a woodland census. The Institute of Terrestrial Ecology (ITE) has developed a land classification system which has been used in a Government funded energy study and, in conjunction with the Highland Regional Council, to monitor land use change. Consideration is being given by the Agricultural Departments to the spatial referencing of farm based data although financial constraints have limited progress.

2.23 Further application of Geographic Information Systems in this area will depend substantially on Government. The increasing requirement for Government Departments to assess the effects of the Common Agricultural

Fig. 2.3 Relationship between a contour map and a digital terrain model

(a) Original contour map

(b) Derived digital heights

```
320 322 299 274 286 246 235 229 220 215 209 200 188 147 103  87

314 308 295 289 257 245 250 257 256 233 219 189 180 138  89 107

291 289 271 258 243 249 261 270 255 245 215 179 158 104  96 130

271 259 251 240 244 248 259 263 251 226 193 173 147 109 102 152

242 235 227 226 258 243 232 241 235 210 182 158 132  98 121 176

233 223 212 214 205 200 209 218 207 193 163 126  97 110 162 187

202 196 188 176 168 153 162 182 178 165 119  95 167 198 210 216

137 112  90 104 131 148 154 147 113  95  86 134 179 201 218 222

 95  91 101 133 152 156 138 103  91 102 123 142 168 192 206 227
```

(c) Graphic presentation of a digital terrain model

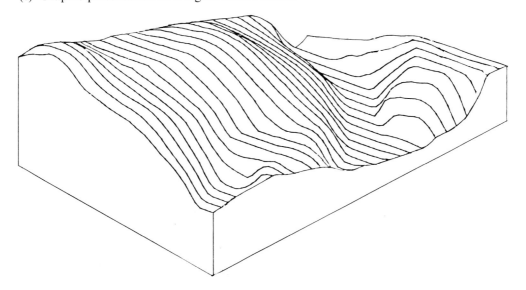

Policy on UK agriculture and to balance the interests of agriculture, conservation, recreation, and the social and economic needs of rural areas could, and we believe should, result in greater Government interest in the use of these systems.

Environmental Monitoring

2.24   There has recently been a marked growth of public interest in the effect of pollution such as acid deposition, the long term effects of changes in atmospheric composition and in radiation fall-out on the environment. There is likely to be an increasing requirement for spatial data to aid analysis, including the use of modelling techniques, in this area. Users include Government Departments, research establishments, and water authorities within the United Kingdom and Europe. Major data sets used include river water quality, national pollutant emissions and climatological and rainfall data. Remotely sensed data can provide information on, for example, thermal discharges from power stations and some aspects of erosion and sedimentation in rivers which can affect water quality.

Epidemiology

2.25   One application area which has recently increased rapidly in size is that of linking health and population data with other data. Following the 1980 Black Report *Inequalities in Health* (Ref. 5), which highlighted the local area variation in the health of the nation's population there has been an increased interest in linking the incidence of disease and other illness in local populations, to factors such as environmental conditions and regional variations in food comsumption. There has also been increased use of linked health and population data for small geographic areas to allocate resources and monitor the effect of local service delivery. This has been encouraged by several reports such as the Korner Committee, Cumberledge and Griffiths reports (Refs. 6, 7 and 8).

2.26   A major difficulty in these types of studies has been the lack of a suitably sized areal unit for matching different data sets. Traditional areas such as wards or health districts can 'hide' local effects. The referencing of population and medical data by unit post code has greatly helped the linking of the two sorts of data, particularly in Scotland where the Census of Population is post code based. The Korner Committee recommended that all medical statistics be referenced by unit post code.

Utilities network management

2.27   The utilities, including gas, water, electricity authorities and British Telecom, are major users of geographic information; its efficient handling is a significant economic consideration for them. They are jointly responsible for 1.65 million kilometres of underground mains with a replacement value of about £117 billion. Their current joint annual workload is estimated by the National Joint Utilities Group (NJUG) to involve 2 million openings in the ground, the notifications for which amount to the equivalent of 50,000 sheets of paper a day. The largest spatially referenced datasets held by each utility relate to its distribution networks, customer records and streetworks notices.

2.28   Some organisations have started digitising all their network records; others are examining the possibilities of doing so. Wessex Water Authority, for instance, has already digitised its entire water supply network. There is a growing recognition of the benefits of linking network data to a large scale digitised map base, as shown by British Gas South Eastern's recent digital mapping trial. In the absence of OS large scale digital maps some organisations − such as the South West Electricity Board − are digitising maps themselves as an interim step.

2.29   Since the 1950 Public Utilities Street Works Act (PUSWA), utilities have been obliged to send notice of intended works to each other and to local authorities. More recently, the Horne Report (Ref. 9) recommended that a central electronic road works register be set up to provide information on roadworks to both utilities and local authorities. We welcome the Government's recent acceptance of that recommendation. In addition to this, the utilities are now actively pursuing methods of using computerised systems to share or exchange data of a broader nature. A number of trials have been conducted, including the Dudley Digital Records Trial where information required by the different utilities is held on one central computer *(Fig. 2.4)* and, more recently, the Taunton Trial in which information is held by the individual utilities and accessed by other utilities through a networked system. Case Study Example 4 outlines computer based developments and trials conducted by Wessex Water Authority, British Gas South Eastern Region, NJUG at Dudley, and the Taunton Joint Utilities Group.

Transport network management

2.30   Applications in this area are centred on the provision and maintenance of transport networks such as roads, railways and canals, and their use for freight and private transport. Users, spanning both public and private sectors, require information on road traffic flows and car ownership levels for monitoring route usage and demand and providing a basis for modelling future demand.

2.31   Central Government (the DTp, Transport & Road Research Laboratory (TRRL) and the Scottish Office) has taken an active part in developing computer assisted information systems for transport and traffic management based on road networks. These include the DTp's Transport Referencing And Mapping System (TRAMS) and the Scottish Office's Computerised Highway Information & Planning System (CHIPS). These or similar systems have been taken up by a large number of local authorities: a few incorporate information on socio-economic and land use data. A more recent system being developed by the DTp is the computerised Network Information System for the management of the road system at strategic and local levels. It will eventually contain trunk road data ranging from road maintenance and construction data to details of traffic accidents and flows. One new application being investigated by the TRRL and private sector is the use of computerised systems to aid vehicle route navigation — see Case Study Example 5.

Property development and investment

2.32   The private sector has a strong interest in access to information on individual properties and land and market developments, for financial investment and property development purposes. Information is required both for specific sites and local areas. For each site or individual property, data describing features such as the area and condition of the site, and the age, size and type of building, are required together with details of ownership and past transactions. Local area information is required on rents and values and on matters relevant to development such as the amount of new floorspace in the pipeline or available for letting and the rate of take-up of new space.

2.33   In recent years the property industry has produced considerably improved information on market rents and values, including the setting up of some computerised property database services, but there are still problems with development data. Much time and money is wasted in disputes between developers and planning authorities because there are no authoritative spatially referenced data on the 'development pipeline'. Information on individual properties is held by the District Valuer, the Land Registries and the local planning authority but access is limited by confidentiality constraints.

Fig. 2.4 National Joint Utilities Group Dudley digital records trial: configuration

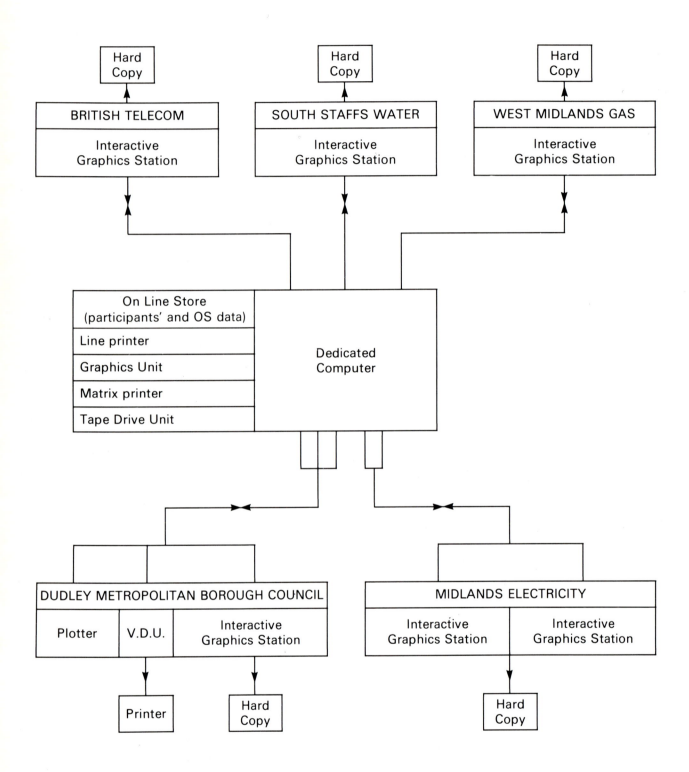

Marketing and business location	2.34 Private companies are increasingly requiring more accurate analyses to give them a competitive edge in locating their businesses — including retailing superstores, offices and warehouses — or for targeting promotions, route scheduling and distribution planning. Such analyses often require the linking of socio-economic and demographic household data with information held in a company's existing customer database: this is done on as small an area basis as possible, whilst preserving confidentiality. A futher application is in combining demographic data with crime and credit information and vehicle accident statistics for small areas in order to establish house and car insurance premiums and credit rating levels.

2.35 Geographic Information Systems can aid analysis of these data sets, using multivariate statistical methods and spatial analytical and simulation models. They can then help to present the results, for example in map form, as names and addresses for targeting or as listings for leaflet distributions — see Case Study Example 6. Use of such systems in this area is still limited but growing, together with an increasing appreciation by companies of the potential value of their existing customer databases.

Civil engineering	2.36 Geotechnical consultants, engineering geologists and civil engineers require information on site conditions, including subsurface properties of materials for the construction of, for example, buildings, roads and dams. Data required includes OS and geological map data, soils and mining records, and borehole logs and site investigation reports. A recent user survey (Ref. 10), conducted by the British Geological Survey (BGS) for the DOE, drew attention to the difficulties many users have experienced in finding out about or gaining access to geological data because of a lack of information on likely sources of data and lack of indexed data holdings. VO evidence to us highlighted the large amount of duplication which exists in the holding of old mine workings records. The BGS's development of the National Geosciences Data Centre, which includes borehole data, should improve the provision of information in this area.

2.37 Relatively few organisations involved in civil engineering activities have developed general purpose computer based information systems. However, for particular highway projects, most county authorities carry out design work, estimation of the volume of earthworks and production of three-dimensional models of the proposed developments using the Modelling & Survey System (MOSS) or a similar package.

Mineral exploitation	2.38 The most sophisticated use of geographic information and Geographic Information Systems in this area is by the oil companies who have made considerable investment in building up computerised exploration and development databases. These bring together a wide range of environmental and physical data on concessions and areas of hydrocarbon potential. They include information on geology, seismic and other geophysical observations and on detailed drill hole information. This information is used to ask 'what if' questions using modelling techniques in both two and three-dimensions. There is well established trading of information between oil companies.

2.39 Companies involved in off-shore work obtain much of the data they do not collect themselves from the Hydrographic Department of the MOD and from the Marine Information Advisory Service of the Natural Environment Research Council. This service provides data and advice on all aspects of physical oceanography, some marine geology and geophysics. One difficulty to be taken into account in marine databases is the recording of the highly dynamic aspects of the phenomena.

Teaching

2.40   For teaching in schools and other academic institutions, access to data across a range of areas is required for demonstration purposes. A few software packages are available for demonstrating to school children the use of information such as that from the Census of Population: the BBC Domesday project provides a low cost database and enquiry system using laser disk storage. That aside, however, there is little interactive demonstration software available for self-help/distance learning or tutorials in academic institutes or elsewhere.

World databases

2.41   UK users and suppliers of geographic information and related products may be involved in applications requiring world or European data. The clearest example of this is in the environmental monitoring area. Increasing interest is being shown in the monitoring and modelling of the Earth's biophysical systems. These systems, which include the stability of the ozone layer because of its control of harmful ultra violet radiation and the carbon dioxide cycle because of its effect on global temperatures, are essential for the continuing habitability of the Earth.

2.42   Important sources of data for these investigations are satellite borne sensors integrated with ground based sampling. The most notable property of the relevant data is that they are extremely dynamic. Complete global data for some variables are provided on a daily basis or even more frequently. There are consequently major data handling problems requiring the most powerful computers for solution.

2.43   Considerable effort is being expended to develop tools which can predict the future behaviour of biophysical systems and to provide the necessary databases for such investigations. The International Council of Scientific Unions has recently established the International Geosphere Biosphere Programme to study global change. To support this programme it has been proposed that a World Digital Database for Environmental Sciences (WDDES) should be developed. The United Nations Environment Programme has established a Global Resources Information Database (GRID) as a pilot project within its Global Environmental Monitoring System.

2.44   Another area of work on a world scale is the World Climate Research Programme, sponsored by the World Meteorological Organisation. This programme includes a study of the inter-relationships between climate and the oceans for which it requires access to world marine databases and meteorological data.

2.45   One notable world nautical database is the Admiralty Chart series produced by the UK Government's Hydrographic Department of the Ministry of Defence. There is growing demand for this series to be in digital form, for civil and military applications.

2.46   A study of the environment is being pursued at an European level via the CORINE project (Coordinated Information on the European Environment). The aim of this European Community (EC) funded project is to produce a spatially comprehensive, compatible and integrated database of all environmental descriptors which have relevance to EC policy.

Overseas applications

2.47 Although our review concentrated primarily on United Kingdom use of geographic information and of Geographic Information Systems, we sought evidence of overseas applications in order to learn from them. The rate and area of growth in applications in particular countries has varied according to local factors such as the quality and availability of geographic information, the general level of technology, and the immediacy of resource management problems. In most countries, the requirements (both statutory and non-statutory) of government at different levels have played a significant part in stimulating the use of these systems.

2.48 In North America rural resource management — particularly forestry — has led the field in the use of Geographic Information Systems. With more than a third of land in public ownership, the national governments led the quest to seek more efficient and timely ways to handle and, analyse geographic information. Existing small scale paper maps were unreliable and lower cost computers with interactive graphics capabilities provided the basis of an attractive alternative. Two early systems to be set up were the Canadian Geographic Information and Canadian Soil Information Systems. A further stimulus to development was the growing use of these systems for the administration of local government taxes.

2.49 The demand for Geographic Information Systems has also continued to grow within government for other applications, particularly defence, and now arises from other public bodies and commerical agencies managing resources themselves. The US Bureau of Census, the major socio-economic data collectors in the US, and the US Geological Survey, the main topographic mapping agency, are currently cooperating to digitise features taken from 1:100,000 scale maps to be used to define the 1990 Census tract (basic collection unit) boundaries. The Census tract information can be aggregated and tabulated by ZIP (postal) code.

2.50 North America is now clearly the leading exponent of geographic information applications and suppliers of the associated Geographic Information Systems. A review of North American experience of these systems, prepared for us by Dr R.F. Tomlinson and his colleagues, is attached as Appendix 6. Case Study Example 7 describes some specific North American applications.

2.51 In contrast to North America, the initial driving force in the application of geographic information technology in many European countries has been the establishment of cadastral information systems. The Swedish Government is in the process of replacing manual systems for property and land registration with a computerised one to save time and money. Denmark, Finland and Norway are also establishing national land registration schemes. The Netherlands Government is introducing a comprehensive land and property information system, with considerable staff savings, to encompass central and local government responsibilities in this area. We understand that computerisation of land registration in West Germany is complete and the country is now examining its digital mapping requirements.

2.52 On a city scale, both Zurich and Vienna city authorities are building up large computerised property and land databases for local planning and administration purposes. The Council of Vienna's database, started in the early 1970's, now includes details of all the utilities' networks, hospitals and medical services and its own properties; it can be accessed from 2,700 work stations.

2.53 The European countries have been slower in developing the use of Geographic Information Systems outside the cadastral area. In Denmark, however, the Ministry of Agriculture's Bureau of Land Data is computerising its data system: data from the computerised pedological soil classification systems have been widely used in physical and agricultural planning. The Netherlands is developing a coordinated programme of research on environmental data handling.

2.54 Cadastral based Geographic Information Systems have also predominated in other countries outside Europe and North America. In Australia, the application of such systems to the land registration and cadastral surveying and mapping area was originally at the instigation of State Governments, although other levels of government are now involved. In Japan, the constraints imposed by the country's mountainous terrain, location in an earthquake zone and high level of settlement makes land a particularly significant resource. A systematic land information system has been formulated in Japan by private and public sectors, coordinated over the last decade via the National Land Agency.

2.55 There is also an increasing demand for more sophisticated spatial data analysis for developing countries. This demand is coming from institutions such as the World Bank, which invests a high proportion of its funds in agricultural projects, and by governments of developing countries wishing to determine regional development policies. There has been a growth in the application of Geographic Information Systems for these purposes, particularly in the South-East Asian countries.

Potential for growth 2.56 From our review of the state of existing applications in the United Kingdom and consideration of overseas experience, we believe the application of Geographic Information Systems is still at an early stage of development. North American experience suggests that we will see an evolution in the use of spatial data handling techniques from an inventory tool which replicates an existing paper records system to a comprehensive management tool which provides a new aid to directing resources. The latter use, requiring a corporate approach to data management, has been developed very little to date in the United Kingdom. The utilities have come the closest to it in their current development of digital mapping and facilities management systems.

2.57 We believe this situation will change rapidly. In most application areas there are signs of significant future growth in the use of Geographic Information Systems. North American and UK suppliers of hardware and software predict a steep growth in demand for their products over the next two or three years.

Information systems for planning and managing public services

A: Department of the Environment, planning appeals information system (relevant case records)

1   The planning appeals information system, or Relevant Case Records System (RCR), is used within the Department of the Environment (DOE) to brief planning inspectors on previous relevant cases related to those currently under consideration. The system is also used by DOE Regional offices and planning policy divisions to monitor trends, provide information on particular themes or issues and to provide answers to queries.

2   The RCR holds on-line over 130,000 case records, covering the whole of England. Initial information on cases is provided by both appellants and the relevant local authorities. Other details such as grid references, digitised policy area boundaries, issues and decisions are generated internally. A digital mapping software package — GIMMS is used to search records to determine whether a case falls within a policy area such as a Green Belt or National Park, and controls production of the plotting of appeal sites.

3   The RCR was one of the first computer-based information systems within the DOE's Planning Command. It was set up in April 1984 to replace the manual Land Use Records system and to provide a more efficient and wider ranging service. The system was developed in-house to a short timetable of 8-9 months for design, development and implementation. Estimated development costs of £350,000-£400,000 were expected to be recovered within two years, largely through reductions in staff and accommodation.

4   DOE staff gained valuable experience from the development, implementation and operation of the RCR system, particularly in the following areas:

Data preparation: Difficulties were experienced in converting some 60,000 paper records into digital form, the task needing significantly more staff resources than had been expected. Particular problems in the implementation and early operation of the RCR were caused by the lack of an agreed standard set of local authority and case-type codes within the Department.

Software: Initially system response times were poorer than expected and computer running costs were higher than anticipated. These problems were related to the software used to develop the system. The software was chosen because of its relatively rapid implementation and lower cost of development, both of which were necessary for RCR given the short timescale. System performance was later substantially improved by means of additional software development.

5   A feasibility study is currently being planned to develop a casework system to automate the handling of planning appeals. The existing RCR and two other systems providing statistics on appeals and the management of inspectors' work programmes are likely to be incorporated within the new system.

B: National on-line manpower information service

1   The Manpower Services Commission's National On-line Manpower Information Service (NOMIS) is a Geographic Information System providing information and mapping of employment, unemployment and population data over a wide range of spatial levels. The system is designed to store and analyse a range of data available for small areas and to make them available to a wide variety of users. The service has been operational since 1982 and is based at the University of Durham's computer centre.

2   The service provides four basic sets of information:

Employees in employment: aggregate data for local areas categorised by males/females; full-time/part-time and by Standard Industrial Classification.

Unemployment: totals for local areas categorised by males/females and school-leavers, age and duration bands.

Vacancies, placings and durations: Industrial analysis of vacancies and placings by occupation.

Population: Tables can be provided from the standard small area statistics of the 1981 Census of Population and population projections for single-year bands.

It is important to note, however, that both MSC and the University of Durham are committed to further enhancements of the systems via the addition of new datasets (a recent example is a migration dataseries) and analytical features such as the development of a spread-sheet capability.

3 NOMIS has a comprehensive coverage of unemployment data. For ward level, and aggregates thereof, it includes all unemployment data regularly published at national or other levels. Full details are retained back to 1978 with summary series of data back to 1972. Most of the datasets are available at a wide range of geographic scales based on the Department of Employment Group's adoption of the ward area as the primary statistical building block. The NOMIS system is used by about 140 registered users throughout the Group, other Government Departments, academics engaged on research projects, local authorities and some commercial organisations. The Computer Centre uses British Telecom's Packet Switched System to provide rapid and noise-free access to the system by these on-line users.

4 The NOMIS database is regularly updated using data gathered and maintained by the Department of Employment and the Manpower Services Commission. Tapes are sent on a regular basis from the Department of Employment Computer Centre and are validated, compacted and added to NOMIS usually within 24 hours. The size of the dataset is reduced by data compression techniques to about a third of the original.

5 Using a simple set of commands, the users can extract, sort, sift and statistically analyse the data to attain the desired results. Output is usually obtained as hard copy but there are options to download data to a local computer for further processing such as the inclusion of data within text created by a word-processing software package. In addition the speed of response and relatively low cost of the system allows on-line search/edit processing to be undertaken by any individual wishing to pursue an investigation whilst on-line. Percentage changes can be quickly calculated between any pair of dates and compared with national average figures for particular industrial classifications. Data can also be presented in map form based on digitised map files of the wards of the whole country. From these building blocks, other areas (parliamentary constituencies, regions, counties, Travel To Work Areas) can be built up and high quality cartographic output provided by the GIMMS mapping system.

6 NOMIS has proved a valuable tool in day to day routine activities of the whole network of Manpower Services Commission regional offices, in central decision-making, in academic and local authority use. It represents a cost-effective use of resources to provide an essential information service and is a prime example of a large and operational Geographic Information System. It has proven to be a successful link between government and the university sector, both parties having benefited from such collaboration.

Local authority land and property information systems

A: City of Glasgow District Council

1   In 1979/80 the Council decided to invest in a computer-based system primarily to manage its housing stock, the largest in Europe. The main priorities were to support the provision of:

- Property portfolio information
- Property use and ownership information
- Housing management information

The Corporate Property Database contains details of all properties within the City, both publicly and privately owned. Access to the database is by street address or unique property reference number. The geocode of the centre point of each building is held in the database to allow the geographic distribution of property-related data to be plotted. The database is used for housing management, property searches and spatial sorting of information using digitised local boundaries.

2   The Council arranged with the Ordnance Survey for the provision of digital map data of the City in a project that will be completed by the end of 1987. The digital mapping will be supplemented with the Council's operational and administrative boundaries and will be linked with the property database to enable property/land attributes to be associated with the maps.

3   Benefits arising from the computer-based system, incorporating digital map data, are foreseen as:

Drawing office benefits: the Council is a large user and producer of paper maps for local plans, strategic planning, development control, building applications, housing, estate management and maintenance. About £300,000 − £350,000 per year is spent on the paper maps system. Digital mapping should enable updating and production of maps, at any scale, to be carried out more cost effectively.

Added value benefits: the digital database can be merged with other databases to create a flexible corporate information system.

Time saved: the Property database has already cut the time (typically 10 minutes) required to find an address for a property search.

B: Cheshire County Council

1   Digital mapping was introduced into the County Council's Planning Department in 1976 with the aims of speeding up cartographical services and aiding the extraction of spatially encoded information from large datasets. The Department holds grid referenced datasets on several topics such as planning applications, industrial premises and development sites. The boundaries of planning areas, such as Green Belts, are also held in digital form. Often these datasets need to be searched to extract information by areas of interest, for example, to compile a list of planning permissions for industrial use in the Green Belt in the last five years. Following the completion of the data processing tasks, the system uses hard copy devices to display the relevant information as maps, plans or graphs, in the form most useful for the ultimate users. In particular, hard copy maps can be overprinted with derived information.

2   The range and nature of the information products derived from the system are illustrated by the following examples:

1981 Census of Population. The statistics were mapped for the 202 wards in Cheshire thus providing a useful illustration of the data. This has been a strong factor in increasing the range of users within the Council and has contributed to the higher degree of analysis of the statistics as compared to previous Censuses.

Areas of family stress. Sixteen indicators of family stress − such as the number of free school meals, probation orders, youth unemployment rates − were collated for each ward within the County. Maps shaded by ward were plotted for each indicator and a summary map showing the ten most heavily stressed areas was produced. This improved the understanding of the varying needs throughout the County and contributed to the more efficient allocation of resources.

Ecological Data Base. The Cheshire Conservation Trust surveyed the County area for sites of ecological value. Some 10,000 sites were classified into 20 categories. The resulting digitised information has been used in Public Inquiries in the form of maps and charts showing a breakdown of the various classifications.

3   The use of digital mapping is regarded as a step-by-step development activity to gain experience in a new field of Information Technology and to make the County Council aware of the potential benefits whilst at the same time solving problems and generating useful products. A feasibility study is now underway to identify the corporate uses of digital mapping and to establish whether other organisations might be prospective partners.

C: Dudley Metropolitan Borough Council

1   In 1977/78 Dudley became the first authority to implement the Local Authority Management Information System (LAMIS) on an ICL 2900 series mainframe computer. This followed the earlier development of LAMIS sponsored by ICL, the Department of Trade and Industry and Leeds Corporation for the ICL 1900 computer range.

2   In addition to improving efficiency by computerisation of traditional functions, the aim of 2900 LAMIS is to provide a shareable pool of land and property information. This is captured from the Departmental systems and linked to a central file which contains every current property in the Borough. At Dudley, the systems linked to the central file are Planning Application and Building Inspections. A Property Terrier system is under development.

3   Linkage is either by cross referencing or by spatial comparison of different boundary types. Dudley acquired a complete set of central file boundaries as a result of a Dudley/Ordnance Survey research project, which led to the derivation of boundaries from digital map data. Dudley have added the boundaries of wards, zones and areas of special interest. The enquiry facilities produce listings, statistics and plots which obviate many time-consuming manual searches both on sites or in filing rooms.

4   As a spin-off from the research project, Dudley also hold a complete coverage of digital map data which enhances the LAMIS outputs as well as improving on the Corporation's traditional map stocks.

5   The Dudley system is an example of a corporate approach to the use of information in an organisation in order to produce an integrated set of data files serving the corporate needs of the Council to achieve greater efficiency in the use of resources and savings in administrative costs. The main uses of the database are, in brief:

Planning and building applications. All applications are monitored and a historic record, back to 1947, is being created. An online system gives histories of all sites.

Building inspections. The various stages of building are monitored and statistics are provided for government returns.

General enquiry system. Many enquiries, both textual and spatial, are made. Examples include: provision of base documents for Planning and Environmental Health projects; listings of firms by type or location used by Careers Officers, and firms who wish to trade in the Borough; provision of statistics to refuse collection division, to give a basis for work planning and assessment of wages.

Property Terrier. This system is under development and will record purchases, and sales of land. Corporation ownership information will be digitised and thus linked to the Central and Planning systems.

Land use and rural resource management

A: Forestry Commission

1 The Forestry Commission is responsible for establishing and managing forests in its ownership. As Forest Authority it is responsible for the administration of grant aid schemes to the private forestry sector and for consultation with other statutory interests in the countryside. It also carries out woodland censuses in Great Britain.

2 The Commission maintains two Geographic Information Systems: one is concerned with the management of its own woodlands and the other relates to the woodland census.

Management of Commission's own woodlands – Sub-Compartment Data base (SCDB)

For management purposes, the Commission's woodlands are divided into compartments. These are the smallest permanent management units used by the Commission and have boundaries based on permanent features, such as roads and rivers. They vary in size from about 10 to 50 hectares.

Compartments are usually further divided into sub-compartments, the boundaries of which can vary through time and depend on the composition of the woodland. The Sub-Compartment Data Base (SCDB) contains basic information such as area, species, age and yield class and at present is used mainly for production forecasting. The addition of National Grid coordinates for each compartment centroid is now largely completed and, together with digitised boundaries of such features as districts and national parks, will aid management planning in the collation and presentation of statistics of area and yield.

The Forestry Commission's data base is updated for major change annually and crop detail is reassessed on an approximate 12-15 year cycle.

Woodland census

Census data for privately owned woodlands (which in 1980 amounted to some 1.2 million hectares) are collected at approximately 15 year intervals. About 45% of the area at that time was grant-aided, mainly through the dedication schemes which provided a comprehensive record of woodland with statutory review and updating. Information was extracted from these records and transferred to a data base. As a pre-requisite for planning the sampling strategies for the 1980 census, the boundaries of all private woodlands not in receipt of grant aid were digitised from Ordnance Survey 1:50,000 maps. Sample woodlands were then selected at random from the woodland size classes, boundary detail updated from recent aerial photography, and a proportion of these woods then visited to obtain detailed crop measurements. The method thus combined grid referencing from maps with air photography and ground survey.

The subsequent withdrawal of the dedication schemes to new entrants and the introduction of alternative grant aid schemes mean that the existing private woodland data base cannot be updated, and the next woodland census due probably in the mid 1990's will consequently be unable to use these records.

B: Institute of Terrestrial Ecology: Land classification system

1 Whilst the environment of Britain generally varies imperceptibly from one area to another, the objective of a land classification system is to produce classes that match the patterns that are present by helping to define the relevant boundaries. The more appropriate the analysis and data used, the better the classes will fit the natural patterns and the more efficiently they can be used as strata for sampling ecological parameters.

2 The land classification system developed by the Institute of Terrestrial Ecology has three phases:
(a) Land classification, based on environmental parameters; (b) Ecological characterisation; (c) Prediction.

Land classification

The classes were determined by analysis of 282 attributes recorded from 1228 1km squares, in a grid at 15 km x 15 km intersections. The data relate to climate, topography, geology and human artefacts and the analysis was used to define 32 land classes. This procedure identifies indicators which may then be used to assign any kilometre square in Britain to its appropriate land class. A total of 6040 squares have been classified in this way. The land classes show well-defined geographical distribution patterns and patterns interpretable in terms of their combinations of environmental features such as altitude, climate, geology and slope.

Ecological characterisation

In this phase field data are added to the initial characteristics of the land classes to improve their definition. The patterns of land use throughout the 1 km square are mapped as well as other ecological information such as hedgerow length and woodland composition.

Prediction

The values from the squares can be used to calculate the mean figures per land class for factors such as area of barley or hedgerow length. As the number of squares in Britain belonging to each land class are known, estimates can be obtained for any factor by summing the land classes for Britain as a whole. Predictive maps can also be made to indicate high concentrations of particular factors.

Applications

3 The Centre for Agricultural Strategy at the University of Reading has applied the ITE system (for the Department of the Environment) to provide a preliminary environmental assessment of possible changes in the farming industry in response to the Common Agricultural Policy and hence to predict the consequences for the rural environment. This analysis was based on the ecological land classification system of attributes defining the physical and biological environments and land use. Using stratified field sampling, realistic estimates were produced of the current allocation of the land surface of Britain into major uses which formed the basis for the model used in the analysis.

4 The ITE system has also been used, in a study funded by the Department of Energy to estimate the area of land that could be available for energy crops in Britain. The resulting land classification was overlain on existing land use information in order to develop an optimal distribution of land for wood energy production.

5 The Highland Regional Council has used the land classification linked with field survey for a number of purposes, most notably in an amenity woodland survey to establish the economic, ecological and landscape significance of woodland. In a survey of the area underlain by the Culm Measures in southwest England, the land classification scheme has been used to provide input to forestry and agricultural modelling in order to estimate wood production and, in particular, the scope for integration of wood production in agriculture.

6 The Department of the Environment is investigating with ITE the ecological consequences of land use change using the Land Classification System as a framework. The main objective is to identify the principal changes taking place in the countryside and their ecological implications. The role of remote sensing in detecting these changes and the conversion of the ecological information into a form readily available to planners will be addressed.

7 In these and other examples, it has been clearly shown that surveying, monitoring and modelling the rural environment can be usefully and efficiently carried out by combining limited amounts of sampled field survey data within the framework of a simple classification of readily available environmental characteristics.

Utilities, digital mapping and facilities management

A: Dudley digital records trial

1 A survey conducted by the National Joint Utilities Group (NJUG) in 1981 indicated that the national cost of mains and plant damage was of the order of £14 million per annum. Recognising the opportunity which digital records offered for exchanging information, NJUG set up the five year Dudley digital records trial in 1982. It is based on the provision by the Ordnance Survey of large scale maps of the area in digital form. The 246 maps at 1:1250 scale covering about 50 sq km are held on a mini-computer and act as a common background for the digitised mains records of all participating utilities. These records are then available for interrogation and up-dating. Utility records are maintained at a 'public' level for access by any other utility or at a 'private' level for the exclusive use of the owning utility.

2 The trial entered its operations phase in 1985 and will be completed in 1987. Software routines have been developed to simplify access to information by staff. One particular feature is the ability to display any small area of a particular map and to produce hard-copy output to give a rapid presentation of all network equipment in the vicinity of a road excavation. The results are shown to an agreed set of standard representations. The last stage of the trial will investigate the use of a computer graphics system to speed up the notification of excavations in accordance with the Public Utilities Street Works Act and assess the financial savings which could result as compared with the previous manual methods.

3 The Dudley Trial is unique in that one databank serves all the utilities and local authorities covering the same geographical area. This could only be achieved by contriving a trial area which represents part of the administrative areas of the different groups participating in the project. The use of such artificial boundaries is unlikely to be acceptable in any future joint records systems. Future work will lie with electronic linking of systems which have been established separately by individual utilities

and local authorities. The common feature will be the Ordnance Survey digital map background with the plant information being interchanged between organisations on the basis of common standards and specifications.

4 The NJUG trial was specifically set up so that the costs and benefits could be evaluated; such an evaluation will be made at the completion of the trial in 1987.

B: The Taunton Joint Utilities trial

1 The Taunton Joint Utilities Group (TJUG) was established in 1983 with the objective of investigating methods for the exchange of records among local utilities. The initial participants in the pilot project were South Western Electricity Board (SWEB), British Gas South Western, Wessex Water and British Telecom; subsequently the local authority (Taunton Deane Borough Council) was also included.

2 The initial trial was based on a 1 sq km area in the centre of Taunton. The Ordnance Survey agreed to support the trial by including the urban area of Taunton in their digital map programme. The utility records for the initial trial area were digitised by SWEB and hard copies for all the records were provided to all utilities. Thus, for the trial area, utilities had detailed knowledge of the location of all underground mains and plant to assist them with their operational activities. While the initial trial was operating satisfactorily, it was evident that the trial area was too small. The Group decided to expand it and to introduce a procedure for the exchange of digital records to reduce the overheads in managing hard copy maps.

3 In July 1985 the trial area was expanded to 6 sq km and the objectives for the expanded trial included:

the monitoring of damage to underground equipment with the longer term objective of assessing the reduction in damage resulting from improved and more timely information;

the examination of the implications and methods for the exchange of digital records.

4   Identical versions of the digital base maps were installed on all the electricity and gas information systems and all utility records were digitised. For electricity and gas, digital records of all utilities data were exchanged using magnetic tape transfers. Other utilities used hard copy maps for the exchange of relevant information. This hybrid operation involving both digital and hard copy records reflects the situation where utilities move into digital records with differing timescales and priorities.

5   The present operations are based on the principle of each utility holding a copy of each others records. These records need to be kept up-to-date and checks are required to ensure that the data are not corrupted. At Taunton, utilities issue a new copy of their records covering a 1 km sq when there is an amendment of the mains and plant details. This procedure has been found to work better than the alternative based on passing details relating to a particular piece of equipment.

6   The trial has demonstrated the need for guidelines on the format for the exchange of data; in particular:

(i)   the format must be transportable across a range of manufacturers' systems and must be independent of the way a particular utility is set up, ie. in terms of resolution and units;

(ii) the large volumes of data associated with utility records, particularly electricity, require the data transfer procedures to be highly efficient.

C: British Gas South Eastern Region

1   This project involved the conversion of all cartographic records into digital form for the South Eastern Region's Crawley District. This area is covered by 240 Ordnance Survey large scale maps and has 46,000 customers. The pilot project started in 1983, went operational in 1985 and has now been completed and evaluated.

2   The major element in the information system is the link between the digital map data and the existing management records database MINE (Management Information in Engineering). The digital map can be used as a graphic aid to search and access MINE and also as a means to support the output of records information. A request for a map of all cast iron gas mains over 5 years old in clay soil would search MINE for the information and then display it on a map through the digital mapping system.

3   The benefits arising from the use of the digital mapping/information system are mainly in cartogrtaphic staff savings and in system management. The system management benefits are related to greater efficiences in:

● Network analysis
● Replacement and maintenance scheduling
● Hazard category mapping
● Route planning (with other Utilities)
● Automated planning and design.

4   The predicted management savings are estimated to be in the same range as the overall cost of the pilot project which was successfully completed in September 1985. The digital mapping system will continue to be used within the Region's Crawley District and further expansion is under consideration.

D: Wessex Water Authority

1   The Wessex Water Authority is responsible for management of water supply, recovery of dirty water and river management over an area of some 10,000 sq km with a population of over 2 million.

Following an early entry into digital mapping in the late 1970's the entire water supply network was digitised by 1985 (see also 2. below). A Geographic Information System has been acquired and development work is now underway to apply the system to meet management and operational needs in a number of areas across its complete sphere of responsibilities.

Water supply. The database is beginning to be used as the definitive plant record, with operational staff updating this record as changes are made to the network. A pre-requisite for this is the availability of digital base maps (see 3. below). Time-dependent data, which was previously recorded manually, relating to bursts and similar incidents, will shortly be added. Management will then be able to review quickly the history of the relevant portion of the network and the replacement programme can be more accurately tuned to the condition of the assets.

Recovery (sewerage). Digitisation of the sewer network in one of the major urban areas (containing approximately one-sixth of the total sewer length) will be complete in mid-1987. Sewers represent a major asset, much of which is now in a state of disrepair. Wessex is drawing up a sewer rehabilitation programme and information on sewer condition is a vital input to this. Digital mapping provides a means for recording all available sewer-related information logically and consistently and also collating and presenting it in a format appropriate to requirements at all levels.

River management. The relationship between abstractions, discharges (including pollution) and the river flow is fundamental to the achievement of the Authority's objectives. River data will show the sampling and extraction points and the upstream/downstream relationship of the network.

2   A significant factor to be considered when implementing digital records is the lengthy and expensive commitment needed to convert the data into digital form. The full benefit cannot be realised until a substantial database has been built up. Wessex implemented a low cost pilot project and data conversion exercise to build up an accurate and consistent database which could be justified irrespective of the outcome of its pilot project.

3   Recognising that digital base maps are required for the cost-effectiveness of the system to be maximised, Wessex has entered into an agreement with Ordnance Survey to obtain maps of its region much sooner than would otherwise have been possible. One aspect of this is that Wessex will become the agent for the sale of digital OS maps within its region outside Bristol, whilst in Bristol itself a bureau will be set up to give on-line access to these maps.

4   A major aspect of Wessex's approach has been that digital mapping, whilst important on its own, should also be integrated into the overall approach to the mainframe databases, for example, so that a user of river mapping information can also access the Water Quality Archive.

Vehicle navigation systems

Benefits and recent developments

1   Inefficiences in route navigation in Britain are equivalent to 6% of all vehicle mileage and cost the nation £2,400 million per year. Vehicle navigation systems could reduce average journey times by 8-12% by helping drivers to find better routes. In London alone these savings would amount to some £100 million per year.

2   Electronic route guidance is a potential application of geographic information which could have a large payback for users. A range of research projects is underway worldwide to indentify the most cost-effective system. The Prometheus project, for example, involves 14 European motor manufacturers cooperating, within the EUREKA framework for research and development, on applying new technology in vehicles, including navigation systems.

3   Within Britain, the recently launched PACE (Plessey Adaptive Compass Equipment) system works out and displays a vehicle's position on an electronic map based on data received from a microprocessor controlled compass and an accurate odometer. The Dutch CARIN system makes innovative use of compact discs for the storage and retrieval of electronic maps to support a vehicle navigation information system.

Autoguide

4   The Department of Transport has issued a discussion document on the proposed Autoguide system which would use a network of roadside beacons, installed at the approaches to all major junctions, to transmit information to and from passing vehicles. The beacons would be controlled from a central computer which would monitor traffic conditions. The driver, having set a destination on the in-vehicle route computer, would receive information on the best routes whenever he approached a major junction. The in-vehicle unit can also be designed to give messages on speed limits, accidents, roadworks and other special traffic conditions.

5   The roadside beacons consist of a receiver/transmitter, a microcomputer, a link to the control centre and a memory unit which contains the signpost information for the junction. Different vehicles can be given different routes — for example to take into account lorry traffic restrictions, or giving routes selected by different criteria (eg quickest or cheapest route taking into account time and distance).

6   The demonstration system, developed by the Transport and Road Research Laboratory, links the route computer and the beacon through small aerials fitted underneath vehicles and inductive loops buried in the road surface. As well as carrying information, the loops can also detect and count all unequipped vehicles, giving extra information on traffic conditions. The major source of useful data, however, will be the information, obtained from equipped vehicles, on the time taken for journeys.

7   The cost of the in-vehicle computer and communications equipment would be about £150 per unit and the cost of equipping junctions within the London area bounded by the M25 would be £15-20 million, with annual running costs of £2-3 million.

Private sector applications of geographic information

I: PRODUCT MARKETING IN THE FINANCIAL SERVICES SECTOR

1   Old-established financial services are facing increased competition from new ventures. Some financial organisations have therefore started to seek more accurate ways of evaluating the performance of their branch offices and assessing their existing and potential share of the market. Some examples of this are outlined below. They are based upon actual assignments undertaken by Pinpoint Analysis Limited, a private sector consultant. In each of these cases, the use of locationally referenced data significantly improved the cost-effective marketing of financial products and services. The main types of information used were:

- 1981 Census of Population, Small Area Statistics

- Financial Research Survey

- Enumeration District Centroid references

- Unit Postcodes and Postal Sectors

A: Identifying which branches to develop

2   The first example is of a financial institution characterised by a large, established customer base, a network of branches used specifically for administrative purposes, and direct mail sales and a quality loan product portfolio. It decided to change its marketing strategy from one based on direct mail selling of competitively priced products to one based on offering personal service. It accordingly intended to refurbish its entire branch network to form personal service branches. The use of geographic information helped it to take a more selective approach to achieving its new strategy.

3   A sample of existing customers' addresses were grid referenced and plotted to identify the pattern of existing customer location. The number of existing customers falling within the digitised boundary of

each branch catchment area was counted and compared with the total number of households, thus indicating the potential market in each catchment area. The results of this analysis showed that the existing customers were unevenly distributed throughout the country, and potential customers were also unevenly distributed between branches.

4   The new strategy could therefore be pursued by identifying 'priority' branches, measured by customer 'potential', where the personal service branch concept could be test marketed. In the longer term, this strategy could be used to develop branch catchment areas to provide a better match between the services provided and the potential customer base.

B: Identifying customers to approach

5   The second example is of a financial institution characterised by a high profile branch network, a large established customer base, limited direct mail sales, and a wide and expanding portfolio of products. Its marketing strategy is based on developing different sales approaches to attract more and different types of customers. One current sales experiment is based on a new 'financial shop' located in a shopping centre in a major town. This was to have been promoted evenly over the whole of the potential catchment area but again a more selective marketing effort could be pursued using analysis of the relevant characteristics of the population.

6   Each postal sector within the catchment area was considered in terms of the number of people and the average characteristics of households contained within it. Financial summary data on usage of various products — such as current accounts — were used to assess potential demand.

7   As a result of this analysis the catchment was

43

considered more closely at the level of Enumeration District. High potential demand areas were clearly defined and names and addresses identified for mailing purposes or for personal sales visits.

II LOCATIONAL ANALYSIS IN THE RETAIL SECTOR

1   Locational analysis is the procedure for evaluating the information relevant to the location of a business or enterprise in order to optimise its position with respect to its customer base. In the case of retail stores, the procedure is used to assess the sales performance of existing stores and to examine the viability of opening new stores. Locational analysis can provide valuable information for management at relatively minor cost.

Methodology and application

2   In a specific application concerning the proposed location of a supermarket in the Oxford area, a computer-based model was built to describe the existing pattern of food shopping expenditure: this model could then be used to support simulations of new store developments. The basic input data for the model were the results of a sample of households regarding their patronage of different supermarkets for food shopping. The sample data, weighted by social class, was summarised at postal district level, the basic spatial unit for the demand-side analysis. The food shopping expenditure statistics at the postal district level, were estimated using Family Expenditure Survey data linked to the 1981 Census of Population data.

3   On the supply side, existing supermarkets were referenced by postcode and national grid references. In addition, the net sales floorspace of each supermarket was included in the model, together with the main road distances between the centre of each postal district and individual supermarkets.

Results

4   The above model was formulated as a micro-computer based software package for use by retail management to compare and contrast the results of various 'what if' simulations. For example, disaggregated data on consumer expenditure patterns, together with information on the number, size and location of competing stores, provided the basis for assessing the potential effects of alterations to product mix, space allocation, rationalisation or expansion of the number of retail stores in the network. The introduction of Electronic Financial Transactions at Point of Sale (EFTPoS) systems will provide additional customer-related information which can be used in such models.

Overseas experience

A: Burnaby (Canada)

1   Burnaby has a population of 135,000 spread over an area of 40 square miles adjacent to Vancouver. Burnaby was one of the first municipalities in Canada to install a computerised system for automated mapping. Some 70-80% of the information used in municipal management was tied to a geographical location, eg planning, public works, police, fire, taxation and a long list of management functions; it was thought that there would be significant advantages in establishing an integrated information system to provide information as a corporate resource.

The project

2   In 1975, the municipality and the local utilities combined to select an automated mapping information management system developed by Synercom. The system includes a mapping graphics and database management package designed specifically for the managment of spatially related information.

3   Burnaby produced a continous digital map of the entire municipality based on aerial photography and ground survey. This high precision cartographic database served as the foundation to which other layers of municipal and utility information – such as water supply networks, census tract data, crime statistics, land parcel and property-specific information – have been added. The database is able to link together information relevant to police and fire services, planning and residential and commercial development. Specific applications are:

Planning: The database was used for providing statistical analyses and for projecting traffic volumes and flows, planning for schools and other requirements based on neighbourhood demographics, housing and industrial development strategies.

Fire service: The system has been used to determine the optimum location of fire stations, with the aim of siting them within a 3 minute response radius from any location in the community.

Police: The location of crimes, types and times of incidents and other information has been collated to establish patterns of criminal activity.

Public works: The mapping of water mains according to user specified criteria (eg more than 30 years old, smaller than 6 inch diameter) can be produced quickly and comprehensively.

Costs and benefits

4   The estimated cost of the Burnaby system amounted to $1.5 million of which $400,000 was for hardware and software, $300,000 to establish survey control and $800,000 for staffing, site preparation and miscellaneous costs. Most of these costs have been recovered by Burnaby through the sale of maps to the Provincial Government and income generated by leasing time on the computer to other users. Map maintenance has been reduced from 340 to 40 man days per year, giving a saving of approximately $50,000 per year.

5   The system is seen as part of a wider information network, including facilities such as word processing, electronic mail, spreadsheet processing, and business graphics. Burnaby expects to install approximately 200 peripheral devices, microcomputers, printers and stations over the next several years as access to the system is enhanced.

B: Connecticut (USA)

1   In 1984 the US Geological Survey and the Natural Resources Centre of the Connecticut Department of Environmental Protection agreed to test the use of an automated Geographic Information System as a means of improving the traditional methods of capturing, storing, updating, analysing and displaying mapped natural resources data. The project involved the convertion of existing mapped data into digital format and the development of techniques to apply them to environmental and natural resource programmes.

The project

2   Connecticut chose the ARC/INFO Geographic Information System for test and analysis based on applications within Government agency programmes. The applications were designed to evaluate the available technology for integrating data and for use in research, planning management and regulatory programmes. The applications were:

Industrial site selection model: This project aimed to identify potential industrial sites by evaluating wetland, flooding, water availability, sensitive environment areas, water quality, acreage, the availability of water supply, sewers and transportation.

Public water supply groundwater exploration: This project identified potential water well sites within a 0.5 mile buffer zone surrounding the service area of a water utility. Unsuitable areas were removed from consideration on the basis of land use and land cover, water quality classification, pollution sources, waste receiving streams, existing water supply wells, zoning and geological and hydrological data.

Database generation for 3 dimensional ground water modelling: This project built up a comprehensive model for groundwater based on data concerning the land surface, water table, bedrock elevations, boundaries between bedrock layers, location of streams, basin boundaries and hydraulic conductivity.

Lessons learnt

4   The applications served to demonstrate the potential that Geographic Information Systems offer in planning, management, regulatory and research programmes where large amounts of spatial data are needed for analysis. Repetitive analysis can be performed rapidly. Data can be aggregated, reformatted and processed to meet specific requirements. The USGS and Connecticut are now working to address issues arising from the projects — including the suitability of existing databases, feasibility of dissemination of data to other users and transfer of data between systems.

3   Factors influencing development

3.1 We have concluded that, although the overall use of geographic information technology is currently relatively low, there is considerable potential for growth in a large number of application areas. We believe that the rate and nature of growth will be influenced by four major factors.

1   the cost of adopting the new techniques;

2   the availability of the data in the required form;

3   the development of better and easy to use techniques for handling digitised spatial data;

4   how quickly people become aware of the benefits of geographic information technology and develop the skills to exploit them.

3.2 Underlying these factors are a number of human and organisational aspects which, although not unique to our area of concern, are in our view crucial to the acceptance and use of Geographic Information Systems. In this Chapter we review in turn how each of the four major factors and the human and organisational aspects are likely to influence the development and use of such systems. Section 3 then deals with certain of these matters in more detail.

Costs of adopting the new techniques

Costs of Geographic Information Systems

3.3 Spatial data processing requires a relatively high level of computer processing power: to date, the cost of equipment and software to meet this requirement has been a major factor in installing Geographic Information Systems. However, over the next decade we believe that equipment costs will continue to fall dramatically. In particular, processing power and data storage capacities will be much cheaper and will allow the development of personal computers which can handle much larger amounts of data than is the case today: this should widen the market considerably. Furthermore, with the expansion of the Geographic Information Systems market, we expect to see an increase in availability of more economical 'off the shelf' systems and software packages and less dependence on expensive in-house systems and software development.

Costs of data preparation

3.4 An even more significant cost element in setting up a Geographic Information System is the conversion of data into a form suitable for entry into the new system. A general problem facing many organisations is the high cost of converting their own paper records into digital form *(Fig. 3.1)*. In the engineering field, for example, it can account for 65% to 85% of the total set-up costs for a system. New technology has not yet significantly reduced conversion costs. Most existing records tend to be relatively unstructured and not suitable for currently available automated data capture equipment such as optical character readers. Even where data are already held in digital form,

Fig. 3.1 **Paper record held by an Electricity Board**
Additional electricity board information recorded on an enlarged OS base map

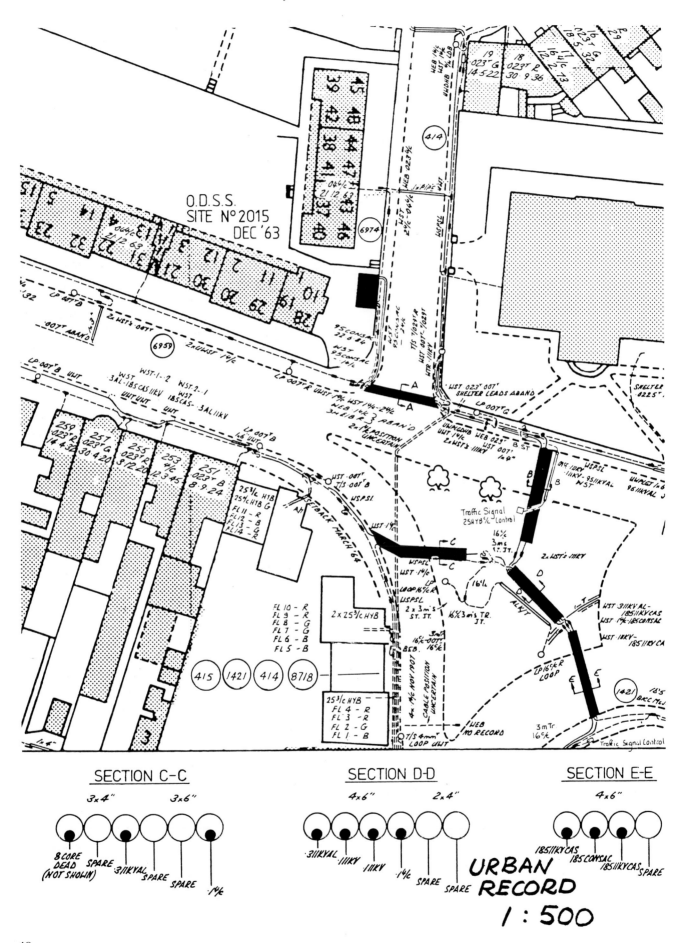

considerable effort is often required to rearrange and code them to permit spatial manipulation.

3.5   Research and development work in this area may result in cheaper methods of capturing structured records in relatively good condition. But the problem will only really be solved in the longer term with the collection of new data directly in digital form, thereby reducing both cost and errors. More discriminate collection of data − including the use of sampling − to meet better defined requirements and sharing of data should also reduce costs (see paras. 3.12 and 3.34). However, in the short term at least, data capture and preparation costs will remain a significant obstacle to the adoption of Geographic Information Systems. In many cases, it will only be the falling costs of equipment and software which will make systems become economically viable.

Availability of data

Ordnance Survey data

3.6   Availability of data in digital form is a prerequisite for using Geographic Information Systems. One key area of data required in digital form by many organisations is the Ordnance Survey's basic scales mapping data. Although digitisation of these data has started, national coverage will not on present predictions be completed until the year 2005. We believe this timescale will act as a significant constraint to the growth in use of geographic information technology in many application areas.

Supply of other publicly held data

3.7   To make the best use of Geographic Information Systems unaggregated spatial data are required. Government, as the major holder of geographic information, often withholds data or releases them only in aggregated form. This is sometimes done to protect confidentiality or to avoid the extra costs of rearranging the data to meet outside users' requirements. It also stems from a lack of awareness of the value of the data, in the right form, to outside users. The Government's recent policy on the collection and handling of information has been to concentrate on meeting its own needs in the cheapest way. Reducing cost and confidentiality constraints to making data available will depend substantially on how much the Government's policy on information provision changes to meet other users' requirements.

3.8   A wider issue relating to the availability of all data sets collected by public sector organisations is the ease with which the user can locate and access them. In the future, spatial data will be held not in the form of one large database but in a series of distributed databases, many of which will be linked by communications networks. Current difficulties in locating and accessing data will grow if these distributed databases are not adequately sign-posted by data suppliers.

Linking data

3.9   For confidentiality or administrative reasons, much spatial data will never be released in disaggregated form. It is currently extremely difficult to link aggregated data because of the variety of spatial areas used as a basis for aggregation and the lack of adequate locational referencing of these areas. Poor data documentation also causes difficulties in linking data: the user needs to know the quality of the various data sets being integrated and the data definitions and classifications used in each. A third difficulty in linking data is the current incompatibility of users' equipment and software.

3.10 We believe this last difficulty will be overcome to a great extent as systems suppliers adopt general Information Technology standards, and inter-connecting equipment and software is created. However, we believe that the first two difficulties − variable spatial units and locational referencing, and inadequate data documentation − will remain significant barriers to linking data unless data suppliers adopt some common standards covering these aspects of providing data. As the largest supplier of national spatial data sets, Government will have an influential role to play in achieving this.

Data collection and updating

3.11 Whilst our terms of reference refer only to data handling, we recognise that it is crucially dependent on how and what data are collected and updated. Data collection is often the largest single cost in the information cycle. At present it appears that data collectors do not in general have sufficient regard for other users when planning the collection and updating of data. The usefulness of data to others can sometimes be increased enormously through small improvements to their collection at relatively little additional cost. For example, the Census Offices' proposal, now abandoned, to collect the 1991 Census on a post code basis would have cost an extra £2 million on top of the basic collection cost of £70 million but would have greatly increased the use and value of the census data. We describe this proposal more fully in para. 6.31. Improving data collection to allow their wider use will again depend largely on the Government's attitude, as the largest spatial data supplier.

3.12 One way of cutting down data collection costs is to use sampling rather than complete surveys. The development of enhanced sampling techniques will be important to ensure that the data collected by sampling can be used in local area analysis. Where full surveys are collected or updated less frequently to save costs, methods for interpolating spatial data for different dates will be required. We again refer to statistical methodology for spatial data handling in para. 3.19.

Remotely sensed data

3.13 We expect the usefulness of remotely sensed data to increase over the next decade. The benefits of the data from the SPOT satellite launched in 1986 − with resolutions of 20m and 10m − and from the recently launched Japanese Marine Observational Satellite and the American Landsats will increasingly be felt, especially if data are made available to users rapidly after acquisition. Two cloud-penetrating radar sensors due to be launched in the early 1990s will increase the effectiveness of remotely sensed data by ensuring their regular supply. Remotely sensed data are likely to be made available under more commercial terms in the future: this may lead to more clearly defined customer requirements and the development of products to meet them.

Better techniques for handling spatial data

Future technological requirements

3.14 Existing Geographic Information Systems and the technology underlying them are still limited in a number of ways. The majority of such systems are still essentially inventory systems, where all that is needed is the spatial referencing of data, such as land parcels or plant, to facilitate spatial retrieval or cartographic reproduction (Fig. 3.2). The emphasis in future Geographic Information Systems will be on manipulating, analysing and modelling spatial data. The ability to access data held in other systems and to integrate it with users' own data, and to use the systems with little expert human assistance will also be required.

50

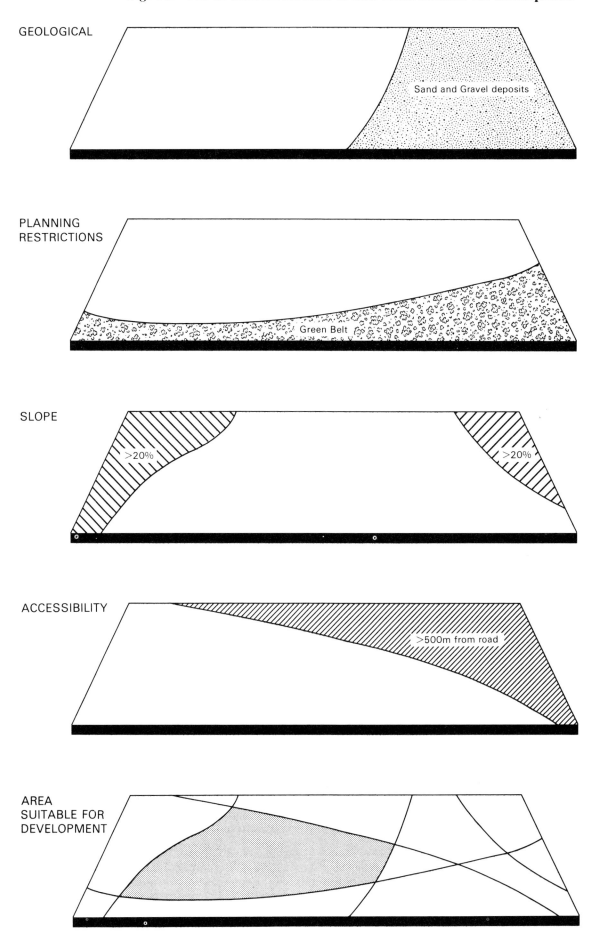

Fig. 3.2 **Use of data inventories to find areas suitable for development**

GEOLOGICAL

Sand and Gravel deposits

PLANNING
RESTRICTIONS

Green Belt

SLOPE

>20%

>20%

ACCESSIBILITY

>500m from road

AREA
SUITABLE FOR
DEVELOPMENT

3.15   To meet these requirements, development of technology is needed in a number of areas. In a few years time, the development of parallel processors and new high density storage – through, for example, optical disc technology – will permit the storage and processing of large volumes of data. However, the full range of Geographic Information System functions may need the development of wholly different database structures to cope with problems arising from the large size of the databases and their effective organisation and management: it will be some time before such databases are developed. Again, high speed telecommunications facilities are available today for special applications, but they will need to be installed nationally before they can be of full advantage to geographic information users.

3.16   We believe that the development of Intelligent Knowledge Based Systems (IKBS) offers considerable potential for greatly improving the flexibility of Geographic Information Systems and the range of questions they can answer. Such systems will have some ability to answer 'what if' questions and they will have a knowledge base. This will lessen reliance on experience and intuition in decision making. It will also provide the capability to carry out spatial modelling which we see as being the major development for the next generation of Geographic Information Systems. It is likely that significant advances will soon be made in this area, initially in relatively simple areas of application.

3.17   Although IKBS should provide the means to increase flexibility of analysis and improve interpretation of results, they will not in themselves improve the human interface with the computer system: 'user friendliness' is a further important area where technological development is required. Further improvements are also required in display technology, particularly for map data. Although it is now possible to produce maps, such as thematic maps or interactive map displays, quickly and cheaply for specific purposes their quality may be an obstacle to their use for some years ahead.

3.18   To provide digital data for Geographic Information Systems, quicker and cheaper ways of capturing paper records, particularly OS maps, are required. With the increased usefulness, and hence take-up, of remotely sensed data users are likely to require Geographic Information Systems which can handle raster and vector data as standard inputs. To meet this requirement, better techniques are needed to integrate the two sorts of data.

Development of statistical methodology

3.19   Accessing and combining different databases will increase the need for the application of sound statistical methods for testing the spatial validity of the resultant ouptut. Development and application of spatial statistical procedures are needed to aid data collection (para. 3.12), processing and analysis.

Research and development

3.20   Some of the technologies needed to improve Geographic Information Systems will come from the adoption of general developments occurring in the broader Information Technology field. Other technologies, and the need to improve statistical methodology, are unique to the geographic information field. Adapting or devising new technologies and achieving their exploitation by users will require a significant amount of collaborative research and development (R&D) between academic researchers, systems suppliers and users themselves.

3.21   It is in the development of innovative products that UK firms may best be able to gain a competitive edge over other suppliers of Geographic

Information Systems. Their success in developing new products will depend on how much effort they put into R&D, collaboration with academic institutions to 'pull through' basic research, their knowledge of users' requirements, and the amount of public R&D funds they receive in relation to overseas competitors.

Increasing awareness and human skills

Awareness
3.22 Adoption of new techniques does not automatically follow from awareness of them; human and organisational difficulties may hinder take-up. Nevertheless, we consider awareness to be a vital first step. Several recent Government and other reports (Refs. 11 and 12) have highlighted the relatively poor level of awareness in the United Kingdom of the opportunities afforded by Information Technology. Our own review suggests that many organisations – from Government Departments to small private sector firms – who might have been expected to have adopted Geographic Information Systems, have not yet done so. This seems to occur because they are not aware of or have little experience of the new uses and benefits of such systems, of their strengths and weaknesses, and of data sources and the scope for data sharing.

3.23 Organisations are more likely to be receptive to geographic information technology if they have already decided to make more use of Information Technology. The rate of take-up of Geographic Information Systems will therefore be affected by measures taken, both within and outside the general education system, to increase awareness of Information Technology in general.

Human skills
3.24 The development, adoption and use of Geographic Information Systems requires personnel at many levels of skills. Personnel also need to know how the information provided by these systems can be used in new ways. Many personnel will be able to acquire some of the necessary skills through work experience. However, in most cases other forms of training are essential to prepare for or complement work experience. Evidence put to us indicates a shortage of trained personnel in this area and particularly of highly skilled people. As the demand for Geographic Information Systems grows, this shortage is likely to increase unless appropriate education and training opportunities are developed.

Human and organisational problems
3.25 There is much evidence that the major difficulties in successfully using Geographic Information Systems are as much human and organisational as they are technical or due to a lack of data of the right sort. We see three major issues: managing information as a corporate resource, the problems of sharing data – not only within an organisation but, as important, between organisations – and the sort of organisational environment conducive to change.

3.26 We emphasise that these issues are not unique to this field; they are, in essence, no different to the problems of harnessing Information Technology generally.

Managing the information resource
3.27 There appear to have been a substantial number of cases where Geographic Information Systems were introduced into an organisation without achieving the benefits which had been expected. Nevertheless, as we have said (para. 1.14), we regard most of these pioneering endeavours as part of the

critical learning process with a new technology; the important point is to note the lessons.

3.28   The first lesson we note is that organisations should regard information as a corporate resource. It is as well, therefore, when contemplating a major investment in a Geographic Information System, to assess the future information needs of the whole organisation, the nature of the decisions that are made and, consequently, what information is required. This last assessment should cover the 'one-off' digitising of existing manual records and the future collection of information in digital form. Both require an estimation of need in terms of detail, quality, frequency and timing. Pilot projects and functional requirements studies can provide effective methods of assessing requirements and of educating users, particularly in the sharing of data.

3.29   Second, advocating a corporate strategy for information is not the same as saying that information should be centrally held. One of the lessons from experience is that the major user of a data set or database should have a long term stake in holding the information; only in this way will he or she have confidence in the data and keep it up-to-date.

3.30   A third lesson is the need for a realistic appraisal of the costs and benefits of investing in new systems. It is important, and relatively easy, to assess the direct cost savings. But it is equally important to assess the benefits which will arise from better decision-making and resource management. These are usually difficult to quantify; doing this properly is mainly a matter of experience and judgement. It is also a good discipline to monitor the costs and use of data once a system has been introduced, so as to provide lessons for future developments.

Sharing data   3.31   Many organisations run separate systems to meet their information requirements for different functions, when they might achieve considerable savings and benefits by sharing the data within the organisation. ICL estimate that typically 60% of data in local authorities could be used for more than one function. In the utilities, the legal requirement to exchange or share data has given them a significant incentive to review data requirements on a joint basis. Local authorities have a wider and more diverse range of services to provide than do the utilities, making the benefits of a corporate approach to data sharing seem more diffuse. However, a local authority might run a transport information system and a social policy planning system with no link between the two, yet both use socio-economic data.

3.32   So far, responsibility for the introduction of Geographic Information Systems into local authorities appears to have been restricted to one department, often the planning department. A single department may not have the resources to develop an application which will be of benefit to other departments although this approach does have the merit of keeping the development of systems directly in the hands of a user department. We found evidence that a few authorities, such as Strathclyde Regional Council and Wirral District Council, were taking a more corporate approach and had set up special coordinating working parties or units. Some local authorities such as the Borough of South Tyneside, have adopted a corporate spatial data strategy as part of an overall information services strategy.

3.33   The sharing of data between local authorities encounters similar problems and merits. In that connection, we were struck by the mutual benefit arising from the joint operation and funding of the Merseyside Address Referencing System (MARS) by a number of public sector agencies. MARS is described more fully in Appendix 5.

3.34   The duplication of data holdings may be due to some degree to a reluctance on the part of data owners or holders to share data because they fear its misuse through misinterpretation once removed from their hands. We understand this view but consider that it may be overcome through adequate documentation and better knowledge of how data can be reliably used. We do not, on the other hand, understand the attitude of those who seem to adopt the narrow attitude that a data holding is in a sense 'their data' and tend to think only of their own needs. This is particularly unfortunate when the information has been collected at public expense.

3.35   Duplication may also arise from a lack of appreciation of the value of data. We found that many organisations take the provision of information, particularly by Government, for granted and have little appreciation of its true value as a key corporate resource.

The organisational environment

3.36   We judge from our review of evidence that the successful introduction of Geographic Information Systems in an organisation is most likely where:

1   geographic information is essential to the efficient operation of that organisation;

2   the organisation can afford some experimental work and trials;

3   there is a corporate approach to geographic information and a tradition of sharing and exchanging information;

4   there is a tradition of a multidisciplinary approach;

5   there is strong leadership and enthusiasm from the top of the organisation, with a group of enthusiasts at the working level; and

6   there is some experience of and commitment to Information Technology and the use of existing databases in digital form.

Realising the benefits

3.37   In our review of the four major factors affecting the future use of geographic information and its handling by computer we have identified a number of significant issues. How and when the benefits of greater use of Geographic Information Systems can be achieved in the United Kingdom will be considerably affected by these issues, which are as follows:

1   how much the supply of spatial data is improved — particularly with regard to the supply of *OS digital map data;*

2   the *availability of other Government data;*

3   the setting of standards to facilitate the *linking of data;*

4   how much effort is put into developing the methodology, technology and products for handling geographic information, through *research and development;*

5   how quickly the general level of *awareness* of the benefits of Geographic Information Systems is raised;

6   how much investment is made in *education and training* of personnel; and

7   *human and organisational attitudes* and approach to the use of information.

3.38 We examine, in Section III, what needs to be done in respect of the first six of these issues. Human and organisational attitudes, a problem widely encountered in relation to the broad field of Information Technology, is a fundamental influence on the take-up of Geographic Information Systems.

SECTION III

Removing the barriers

4 Digital topographic mapping

Introduction 4.1 Many of the UK organisations giving evidence to us stressed the importance of up-to-date mapping as a base for the development of Geographic Information Systems. In all land-related cases this referred to the topographic mapping of the Ordnance Survey. For computer-based Geographic Information Systems, this topographic mapping is usually required in digital form. Topographic information in digital form is required for four main purposes:

1 To enable users to relate their own digital data to topographic features (e.g. a gas main to the edge of a building).

2 To aid selection of other data for analysis and presentation (e.g. all addresses within 50m of a water hydrant or the shortest road route between 2 points).

3 To facilitate the display of topographic information in a way which is most appropriate (e.g. by creating a digital terrain model or by suppressing unwanted detail from a map display).

4 To facilitate the manipulation or linkage of other data sets.

4.2 The responsibility for topographic mapping of Great Britain lies with the Ordnance Survey, which we refer to as OS. Northern Ireland has a separate organisation which we refer to as OSNI. The OS is a Government Department, located at Southampton but with more than 150 small offices throughout Great Britain. The Secretary of State for the Environment is responsible for its work. In addition to the maintenance of the national geodetic framework, commercial map publishing, and a number of special tasks in support of central government, the OS's principal remit is to prepare and keep up-to-date the national map series. These currently comprise more than 220,000 separate maps at seven principal scales. Of these the most fundamental are the basic scales (1:1250, 1:2500 and 1:10,000) at which the survey is conducted and which are the largest scales available for any given area; these account for 97% of the total number of maps. Complete national coverage is provided at smaller scales of 1:10,000 (derived), 1:25,000, 1:50,000, 1:250,000 and 1:625,000; these maps are often known as derived scale maps *(Figs. 4.1 and 4.2)*.

4.3 The work of the Ordnance Survey was reviewed in detail by the Ordnance Survey Review Committee (OSRC) in 1979 (Ref. 13) and, to the extent that the work of OS impinges on the subject of our Enquiry, we have tended to rely on the Committee's Report. We have not, for example, concerned ourselves with the content of OS maps, even though the point came up frequently in evidence from particular interests, because the issue was considered by the OSRC and a recommendation made that a continuing review was the responsibility of OS and its Consultative Committees. There are, however, issues affecting the OS where circumstances have changed since 1979, and these we consider.

Fig. 4.1 Ordnance Survey map series

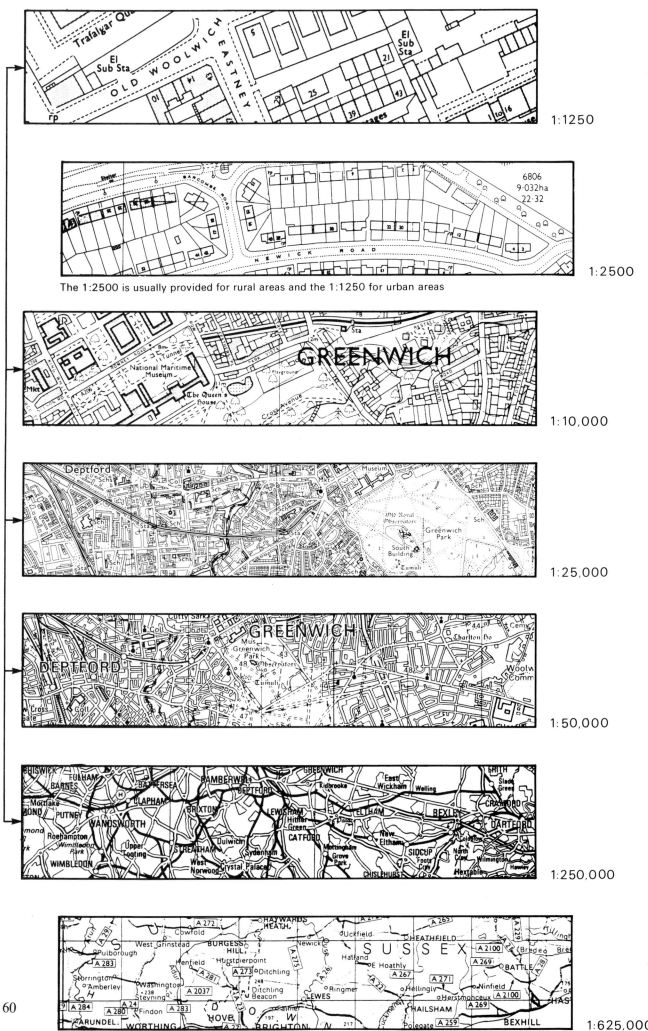

1:1250

1:2500

The 1:2500 is usually provided for rural areas and the 1:1250 for urban areas

1:10,000

1:25,000

1:50,000

1:250,000

1:625,000

London area not mapped at 1:625,000

60

Fig. 4.2 **Links between Ordnance Survey map series**

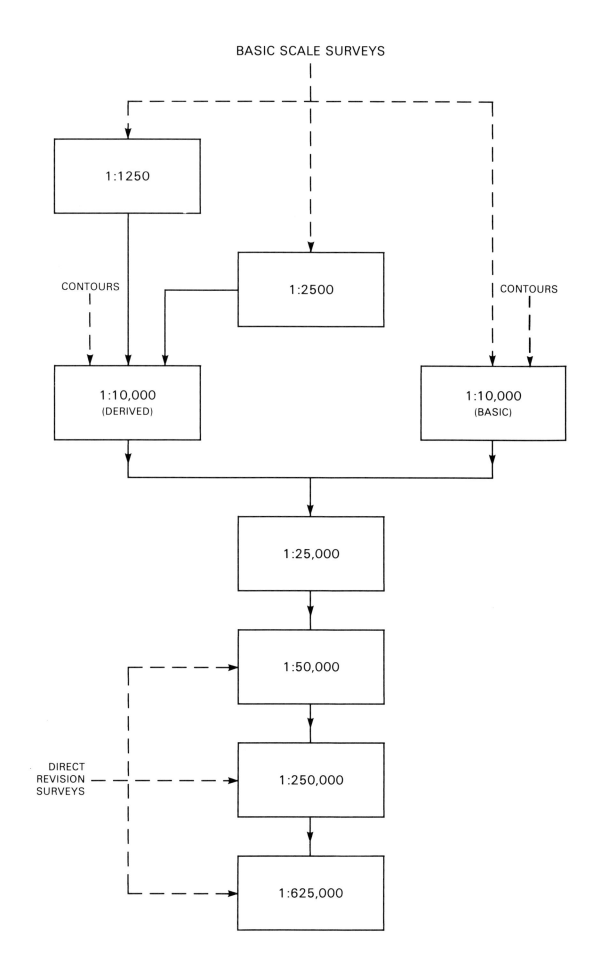

4.4   In this Chapter, we discuss the issues under three headings:

1   Conversion of the existing archive of basic scales maps into digital form.

2   Policies for digital mapping where the basic scales maps have been converted into digital form.

3   Conversion of derived scales mapping into digital form.

It is important to distinguish the one-off task of converting the existing archive from the continuing task of maintaining and keeping up-to-date a digital database once this initial conversion has been completed.

Conversion of the existing archive of basic scales mapping into digital form

4.5   Many users of basic scales mapping expressed their concern about the lack of digital mapping for their areas and about the extended period that will be required at current resource levels for OS to cover the whole country. OS now expects to cover the major urban areas by 1995 and any other areas for which there is demand by 2005. Some utilities have arranged to digitise from OS maps themselves rather than wait for the OS programme.

The current situation

4.6   Great Britain has been fortunate in having a single national series of maps based on a countrywide Ordnance Survey National Grid. OS started its programme of digitising the national map series in 1973. More than 30,000 basic scales map sheets (over 13% of the total) are now available for purchase on computer tape *(Fig. 4.3)*. Some £22m has been spent on the digital mapping programme to date and the Government has already made allocations of £2.5m and £4.55m for 1986-87 and 1987-88 respectively.

4.7   The OSNI has a programme to complete the conversion of Northern Ireland topographic mapping to digital format within 10 years and stresses that its aim is to create an on-line topographic database, holding data in point, line and polygon form together with associated or attribute data, rather than a cartographic databank. This distinction is important in that it highlights the aims of manipulating the topographic data and linking them with other data in a flexible manner, rather than merely using topographic data to produce maps as a background for this other data. We return to this subject, in the context of the OS(GB) programme, in paras. 4.28 to 4.30.

4.8   We commend the work being done by OSNI, both for its approach to a topographic database and for the use of techniques for rapid digitising. We were also pleased to learn of the active cooperation which is being developed with users and potential users of digital systems. The Northern Ireland programme is, of course, smaller than that which confronts OS(GB), but we hope that it will continue to develop along its present lines, in which we see great promise.

4.9   With the current OS digital map, data are held primarily as a set of vectors in the computer which correspond to the lines, points and text of a hard copy map. The vectors are all labelled generally with one of the 160 feature codes such as those indicating fences, buildings, roads, etc. This enables the coded features to be identifed and selected within the computer.

4.10   OS produces its digital maps by manual digitising from the field surveyors' Master Survey Drawing (MSD). The MSD is the up-to-date working document comprising the published map sheet as amended by subsequent survey work. Currently it costs OS on average about £800 to digitise, check and edit each MSD.

Fig. 4.3 **Progress on basic scales digital mapping**

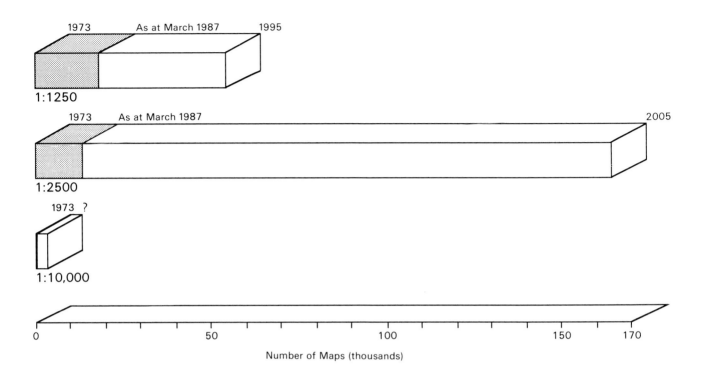

1973   As at March 1987   1995

1:1250

1973   As at March 1987   2005

1:2500

1973 ?

1:10,000

0   50   100   150   170

Number of Maps (thousands)

4.11 The Government has set OS the financial target of meeting 40% of the costs of its large scale mapping activities from sales and royalties by 1989/90 and expects OS to apply this target to large scale digital mapping. In line with this policy, OS raised the selling price of a digital map to £50 from April 1986 and has announced a further price rise to £85 from April 1987.

Users' requirements for more rapid national coverage

4.12 The requirement for comprehensive digital coverage is much the strongest from the utilities who, for operational reasons, say they need national coverage in 5-7 years. But HM Land Registry (HMLR) and the Registers of Scotland could also require full coverage in less than 15-20 years. The evidence on need from local government was more varied. There are some local authorities who would like full topographic digital coverage as soon as possible because they intend to digitise their related records. Other local authorities, particularly highway authorities, want selected features, such as road centre lines, in digital form for specific systems such as highway inventories. Many local authorities, however, cited financial constraints as a reason for not embarking on such digital mapping at the present time.

4.13 We believe that the position has now been reached where there is a real danger of widespread duplication of map digitising by individual agencies. The potential waste of national resources, both in the initial investment in digital maps and also in their on-going maintenance, is considerable. There is also a danger that the exchange and merging of digital data between agencies could become impossible. This is especially important in the case of the utilities, which have a statutory obligation to exchange plant location data with each other. The National Joint Utilities Group (NJUG) emphasised the need for a common digital map background for this purpose and we strongly endorse this view.

4.14 We are also of the opinion that the rate of take-up of OS digital data will be slow until very large areas of the country are digitised. For the main users, individual digital map sheets or even large blocks are of little value unless they embrace the whole of an operational area.

4.15 We therefore believe it to be of the utmost importance to find a solution to this problem. As we explain below, this is not just a matter of the resources put into the digitising programme; it also involves difficult interlocking issues of the techniques used to digitise the maps and the OS map specification. The OS advice, given in evidence, was that national coverage in 5-7 years to the current OS specification could not be achieved even by devoting additional resources to the task. Leaving aside whether either the Government or users would be prepared to find the necessary funds (over £140m), we accept there would be severe operational constraints to achieving such a programme.

Digitising techniques

4.16 We heard much evidence, some highly technical, on the relative advantages and disadvantages of the various methods for digitising a map. These methods include simple and slow manual digitising, as currently practised by OS, and largely automatic systems based on scanning or line-following devices. We were told that automated techniques now available permit the rapid provision of digital map coverage with a limited grouping of feature codes and some reduced cartographic elegance but at a cost very much less than manual digitising using the full range of 160 codes.

The OS digital map specification

4.17 A number of organisations, but particularly NJUG, have suggested that the current OS specification for digital topographic data is too elaborate for their immediate needs and that, by modifying the specification, greater advantage could be obtained from automated digitising. OS however, has to take account of the generality of user needs and wishes to ensure that a satisfactory specification for the national archive of survey data is maintained. It is wary of proposals which it thinks might restrict access to the full range of data and prevent mapping of a satisfactory nature and other applications in the future.

4.18   The main points of agreement between NJUG and OS are that any revised specification should maintain the current OS standards for line accuracy and that the maps produced should not require significantly more storage on the computer than do current OS digital data. NJUG would be prepared to lose a small amount of the map detail (e.g. buildings with an area less than 6 sq.m.), but OS would not wish to lose or degrade any of the information which a fuller specification would require. However, the main points of difference between NJUG and OS are in relation to the number of feature codes used, the 'tidyness' of linework, the positioning of text and the level of quality control. There are cost penalties attached to achieving the current OS specification on these points so it is important to establish whether the users would prefer quicker and cheaper national coverage with the consequent penalty of a less complete specification.

4.19   At a late stage in our Enquiry, NJUG had agreed with the Land Registries and was seeking agreement with the Local Authority Associations to a revised specification which could be put to the OS as representing the requirements of OS's major customers. These three groups account for the major part of OS income from basic scales mapping at present. In this specification, the lines, points and text on the OS map would be classified into 15 feature codes rather than the 160 feature codes on the existing OS specification. For example, the NJUG specification proposes to group all water features in a single code compared with 9 different codes under the OS specification. Similarly, the NJUG specification calls for a single code for building outlines, compared with 12 under the OS specification. It is claimed by NJUG that this reduced number of feature codes, together with a less demanding attitude to 'tidyness' and less emphasis on quality control, would enable efficient use to be made of automated digitising techniques. This would have the result that the cost of digitising could be reduced from the current OS average of £800 a sheet to around £200.

4.20   The cost advantage claimed for the NJUG specification is so large that it merits urgent investigation to see whether the claims on cost and user acceptability can be substantiated. The important questions are:

1   To what extent would digital maps of, or close to, the NJUG specification meet the requirements of the utilities and other users?

2   How much would such maps cost?

3   Given the answers to 1 and 2, are the savings in cost and time sufficient to outweigh any shortfall in meeting most users' requirements?

4   Could such maps be brought up to any future OS specification at a later date without any significant cost penalty?

4.21   The utilities have done some tests to provide answers on requirements and costs (questions 4.20.1 and 4.20.2). The OS are also conducting tests, to find out from users − including local authorities and OS itself − whether the NJUG specification map meets their requirements. However, these questions about meeting user requirements and costs can only be answered reliably by commissioning digital maps to the NJUG specification and testing them in an operational environment. British Gas proposes to order NJUG digital maps for a part of its South Eastern region and this should provide an opportunity for a thorough evaluation of the NJUG specification. It is important that the other major users should collaborate with British Gas so that they can obtain the NJUG maps and test them in their own working environments. We regard the major users in this context as the utilities, local authorities, Land Registries and the Ordnance Survey itself.

4.22   Individual users can, of course, make their own judgements about the extent to which the cost and time advantages of the NJUG or similar specification map are sufficient to offset its possible limitations (question 4.20.3). However, users will need to know the cost to them of obtaining such a map. The cost to users will depend critically upon the number of users and the arrangements for copyright and royalty payments to OS. It is essential that estimates of the number of users and of the likely cost to users are discussed and, if possible, agreed between OS and the major users.

4.23   OS is now (February 1987) running tests to assess the cost of bringing an NJUG type digital map up to the full OS specification at a later date (question 4.20.4). As part of these tests, OS is examining various amendments to the NJUG specification to see whether these more nearly meet its own and other users' requirements. In this connection, it was stressed to us by OS that its current specification for digital mapping is not immutable and would be changed if, by so doing, OS could meet its customers' requirements in a more cost effective and timely manner.

4.24   Our own enquiries have convinced us that is is essential that OS and the major users of basic scales mapping tackle these problems on a collaborative basis. If individual users, or the OS, pursue their own solutions independently of each other there are risks of duplication, the possibility that users would not be able to exchange map data or link them to other data and the possibility that the financial basis of OS could be undermined. The result would be that users would end up paying more for products which do not meet their requirements particularly well.

4.25   We therefore welcome the initiatives now being taken by OS and the major users together to test the NJUG specification, possible amendments to it, and methods of upgrading the maps at a later stage. We also welcome the proposals from British Gas to test the NJUG specification in a working environment with the active collaboration of OS and we stress that it is essential that other major users also test the maps so that they can evaluate the results for their own purposes.

4.26   During the course of our Enquiry we have formed the impression that there has been insufficient communication between the various parties. It is for this reason we have stressed the need for a collaborative approach. As one facet of this, we believe that there is a need for a joint working group of the OS and the main basic scales digital map users to provide a focus for this collaboration. We see this working group as being additional to the existing Consultative Committees which are not suitable for the close collaborative efforts that will be needed over the next few years. OS should therefore convene a working group with the users and potential users of basic scales digital mapping to meet this need. It should be limited to consideration of basic scales digital mapping and its remit should include the specification(s), quality control, costs, funding arrangements, timetable and plans for upgrading of basic scales digital mapping.

4.27   If something close to the NJUG specification digital map proves acceptable to the main users, could be produced at about the costs claimed by NJUG and could later be brought up to a fuller specification without unacceptable additional costs, there would be strong reasons for OS to switch its digitising resources into a more rapid conversion programme. On these assumptions, if OS resources, either from Government allocations or from revenue earned, are insufficient on their own to fund the rapid conversion programme, then we would expect the major users to meet the extra costs. Whichever way such a programme is carried out, by OS or by other organisations, it is essential that the end result is a unified digital database for the whole of Great Britain, with the OS responsible for ensuring suitable standards of quality and consistency.

Users' requirements for a structured basic scales database

4.28   There are other aspects of the OS digital map database which have been drawn to our attention. A number of organisations mentioned the lack of any means within the computer of relating the various lines depicting a building and land parcel either to each other or to the address of the building. The only means of obtaining these relationships is by looking at a graphic display.

4.29   A digital topographic database should not seek merely to replicate the hard copy map. With digital databases, new features or relationships can become more important than traditional map detail. Examples brought to our attention included the need for road centre lines for describing the highway network in highway maintenance systems and network routing and the need for Grid referencing of properties for property record systems; only the former is available on the current OS digital map (and is not stored as a network) and neither is available on a hard copy map.

4.30   The OS is aware of these problems and opportunities and in 1985 started a fundamental review of the structure, content and future use of the digital topographic archive in their Topographical Database Study. Interim findings from the study are due to be published in Spring 1987 and OS intends to hold a series of seminars to discuss these with a view to reaching conclusions by the end of 1987. Clearly the results from this study could have an important bearing on decisions on the specification for OS digital maps. It will be necessary to consider to what extent users' requirements for an enhanced structure for digital maps are compatible with the more rapid conversion programme that we hope will result from the trials of the NJUG specification.

Timing of decisions on a more rapid conversion programme and a structured database

4.31   From the work already done by NJUG (para. 4.21) and the tests (para. 4.21 and 4.23) currently being undertaken by OS, it should be possible for OS and the major users to come to preliminary conclusions by late Summer 1987 on the viability of a more rapid conversion programme using the NJUG specification or something similar. Results should be available on how far the NJUG specification meets users' requirements, on possible amendment to the specification, on whether the maps so produced could be brought up to a fuller OS specification at a later date and on the costs of these different elements. The aim should be to reach final conclusions no later than December 1987 on both the rapid conversion programme and the database structure for the future full specification.

4.32   The conclusions agreed, if possible, between OS and the major users should form the basis of advice to Ministers about the future digital mapping programme. If agreement cannot be reached between OS and the major users, we believe that independent advice should be made available to Ministers before decisions are taken. Decisions will then need to be taken quickly so that OS and users can make plans. The aim should be to announce the final decisions by April 1988.

4.33   *We accordingly recommend:*

1   *OS and the major current users of OS basic scales mapping should collaborate in full scale tests in operational environments of maps digitised to the specification developed by the NJUG.*

2   *OS should convene a working group with the potential users of basic scales digital mapping to guide the development of an accelerated digital mapping programme. The working group's remit should include the specification(s), quality control, costs, funding arrangements and timetables, for creating both initial basic scales digital mapping and any subsequent upgrading required.*

3   *OS should be prepared to switch its resources into a more rapid digital conversion programme if the NJUG specification, or something similar, can later be brought up to a further fuller specification without incurring unacceptable additional costs.*

4   *Should the more rapid programme result in extra funding being required, this should be found by the main users.*

5   *The more rapid conversion programme should produce a unified digital database for Great Britain with the OS responsible for ensuring suitable standards of quality and consistency.*

6   *The aim should be for conclusions on a more rapid programme and on the future full OS database structure to be reached no later than December 1987, with a decision being taken on the way to proceed by April 1988.*

7   *These conclusions should be agreed between OS and the main users. If OS and the main users do not agree, independent advice should be made available to Ministers before decisions are taken.*

The results of all this work are likely to have a fundamental effect on the future provision of digital mapping for Great Britain: it is important to get the decisions right but also to make them before it is too late to implement the choices made.

OS policy towards map digitising by users and bureaux

4.34   A 5-7 year conversion programme for digital mapping will require very large amounts of digitising to be undertaken by users or by bureaux. Unless the work is all undertaken on contract to the OS, there will be a need for special terms on copyright and royalties to cover such a large national exercise. Such terms, if they are needed, should be promulgated when a decision is taken on the programme so that users and bureaux can have the necessary confidence to invest in staff and equipment.

4.35   We also received evidence to the effect that OS policy towards digitising by users was unduly restrictive and difficult to understand.

4.36   To digitise an OS hard copy map a user, or data supplier on the users' behalf, must:

- obtain permission from OS (which is rarely refused but often takes some weeks to obtain)

- pay OS a royalty of £6.50 a unit if the digitising is not done in-house

- not sell the digitised map to other users, although exceptionally the OS will allow the user to sell the digitised data to a third party for a royalty payment of £25 a unit

- if required, supply OS with a copy of the data for OS use.

The OS is prepared to consider waiving the £6.50 a unit royalty where digitising is undertaken to OS specification and it welcomes any other proposals from potential users or data suppliers. We noted, however, that many users felt that OS policy in this area was not sufficiently well defined for them to use it as a basis for investment decisions.

4.37   The evidence suggests that many users only wish to digitise a few map sheets or, more frequently, particular features of map sheets (such as road centre lines). We believe it is important, in the interests of deriving the full benefits from the use of computer techniques in this area, to accommodate this demand. In particular, those carrying out such digitising should be able to negotiate terms with OS in advance whereby they would be able to sell such data to third parties.

4.38   We also consider that the current incentive of a waiving of OS royalty charges of £6.50 a unit to organisations who are prepared to digitise to OS standards is a singularly ineffective inducement, particularly when coupled with the OS control of sales to third parties and when set against the cost of about £800 of digitising a sheet to OS standards. Whilst we understand that OS would take other factors into account in negotiations, the initial impression is one of discouragement rather than encouragement to the potential digitising organisation.

4.39   *We recommend that:*

1   *OS should give details of the copyright and royalty arrangements for digitisng of maps for a more rapid conversion programme when a decision is taken on the programme so that users and bureaux can have the necessary confidence to invest in staff and equipment.*

2   *OS should publish guidelines explaining the principles on which it makes judgements on copyright and royalty arrangements to apply to new proposals for external digitising.*

3   *External organisations should be able to negotiate terms with OS in advance for the marketing of digital data to third parties. OS should insist that such data is adequately documented as to its quality and limitations.*

4   *OS should offer a much larger incentive to organisations prepared to digitise to OS standards where the resulting product can be of value to OS.*

Policies for digital mapping where the archive has been converted into digital form

4.40   Many organisations expressed concern about the OS charging policies for digital mapping and about the arrangements for updating digital maps. There was also strong demand for OS to be more up-to-date with its survey work. We consider these issues below.

4.41 The introduction of digital mapping introduces new factors into the determination of charging policy for OS mapping:

- the difficulty of measuring or estimating the amount of use when there is no easy proxy like the amount of copying;

- the need for OS to recoup a proportion of the initial cost of converting existing records to digital form;

- the extra value of a digital map compared with a hard copy map;

- the balance between the price of tapes and royalty charges which could influence the level of investment by users.

4.42 In response to this situation, OS brought out proposals in January 1986 for a revised method of calculating copyright licence fees covering both digital and hard copy maps. The essence of the OS proposals was that users would be charged copyright fees based on an annual royalty charge for each map held by the user. For digital map users, the royalty charge for each sheet would be increased by a factor determined by the capability of the user's computer mapping system.

4.43 The utilities thought that computer processing power would be only marginally related to the use of digital mapping, whereas local authorities thought that the number of map sheets held by them would not be a good indicator of use. OS should therefore continue discussions with users to develop a charging basis which, while maintaining OS revenue, is considered by users to be fair.

4.44 At present, the OS only supplies digital data at reduced rates for research specifically on digital mapping; this could be extended to wider research and educational uses, as is already done with traditional products.

4.45 In the digital era, it will in due course be unnecessary and even undesirable to sell digital tapes only as whole map sheets. The possibility of users being able to purchase selected features (e.g. roads) would be of advantage to those who wanted wide coverage of, say, a county or district for individual features but would be deterred by the cost of buying all the data when they might want less than 20% of it. This could help to stimulate sales of digital data and it could provide valuable feedback to OS on what customers really want. It could, however, have implications for the way in which OS organises its digital database and needs to be taken into account when reaching conclusions on the future full database structure (paras. 4.28 to 4.30).

4.46 *We recommend that:*

1 *OS should continue discussions with users to develop a charging basis which, whilst maintaining OS revenue, is considered by users to be fair.*

2 *OS should consider whether it would be to its commercial advantage to encourage the development of digital mapping by reducing charges for digital mapping for wider research and educational uses, as it already does with its traditional products.*

3 *OS, either itself or through agents, should sell the separate elements of its digital mapping at appropriate unit costs as soon as this is commercially viable.*

| Revision of OS digital mapping | 4.47 Many of the 30,000 digital maps held by OS have not been revised since they were digitised. In principle, where there is demand for current data, maps in the OS digital databank are brought up-to-date when there are roughly 50 units of change i.e. on the production of a new microfilm (SIM) update. A user wanting the updated information has to buy a completely new tape at the current selling price of £50. |

4.48 Some users of digital mapping emphasised to us the unsatisfactory nature of the current updating arrangements for digital mapping. OS is investigating means of more frequent updating and alternative ways of supplying the update information. We recognise that such measures may be expensive and must be supported by demand. We noted, however, that the cost of an updated digital tape of the same map sheet is the same as an original one. Not only is this unreasonable, but it is likely to deter users from buying digital update data from OS, leading both to duplicate digitising and loss of revenue to the OS.

4.49 In the longer term, we would expect that the digital field updating methods being tested by OS together with appropriate modifications to the database structure will enable digital users to obtain up-to-date digital data soon after it is surveyed.

4.50 *For the meantime, we recommend that OS should continue to develop a better update service for digital mapping, and should consider introducing some reduction in charges for updated versions of a digital map to purchasers of an earlier version.*

Survey data for new developments

4.51 To some extent, discussion of the collection — as distinct from the handling — of data is outside our remit. However, in the OS context, data collection and digital data handling interact in a number of ways and it is difficult to regard them as independent; for example, the type of data collected influences digital data structure, the cost of data collection for one map sheet affects resources available for the subsequent stages of digital mapping, and digital field updating depends on the prior availability of digital mapping. One particular aspect of this frequently drawn to our attention was the need for earlier survey data for new developments. Those submitting evidence were concerned about the lack of early OS involvement and the perceived duplication of effort when several bodies each did their own separate survey of a new development at the start, during and after its construction. (The current OS remit is for accurate 'as built' information which is often different from the original plans).

4.52 It was suggested by some organisations that the local authority should be the official depository of this early survey information which would be held until OS did the definitive survey. Initial information about new developments could be recorded as a separate level in a computer system.

4.53 *We recommend that local authorities and utilities in Great Britain should agree procedures for the common recording of topographic information about new developments prior to the definitive survey by Ordnance Survey. This infomation should be available to Ordnance Survey as pre-survey intelligence.*

Conversion of derived scales mapping into digital form

4.54 Some users of digital map data require information (such as contours) not provided on the 1:1250 and 1:2500 scale maps; others need to analyse data for large areas but do not require all the detail of those maps; and still others use smaller scale maps as part of their existing work. The Ordnance Survey Review Committee recommended that, whilst a single scale-free digital database for the whole country produced from the basic scales survey was the ideal, practical considerations meant that there would need to be two separate databases, one produced from the basic scales and the other from the 1:50,000 series.

4.55 Since then, OS has concentrated almost entirely upon the digital mapping of the basic scales maps. One experimental map sheet at 1:50,000 scale had been digitised by 1985 and a complete, structured database of the information shown on the two map sheets covering Britain at 1:625,000 scale has also been digitised and is on sale. At the same time, other agencies have digitised substantial components of certain OS maps for their own needs. In particular, the Ministry of Defence is acquiring much of the contour coverage of Britain digitised from the 1:50,000 scale maps for terrain simulation purposes and the Institute of Hydrology has coordinated the digitising of water features from the same maps. OS has recently agreed to market the former data sets. For Northern Ireland, OSNI is creating a small scale topographic database from its 1:50,000 series. Contours and water features are nearly complete and preliminary work on communications is in hand.

4.56 Although OS has carried out or commissioned very little digitising other than from the basic scales maps, they have carried out a user needs study as recommended by the Ordnance Survey Review Committee. This demonstrated considerable interest in products derived from the 1:10,000, 1:25,000 and 1:50,000 scale maps and relatively little for those from still smaller scale maps: about two-thirds of large users claimed they required such data before the end of the 1980's but there was universal concern at the price levels which OS proposed to charge (e.g. £460 for a structured 1:10,000 sheet and £60,000 for a structured 1:50,000 sheet). As a result, OS decided only to proceed with digitising a second 1:50,000 scale map and making it available at low cost for experimental purposes and for further testing of market demand.

4.57 The evidence to us on a topographic database from derived scales mapping was not as voluminous as that on the basic scales database. However, scientific organisations − such as the Natural Environment Research Council (especially the British Geological Survey) and the Royal Society − regard digitising at these derived scales as a high priority. We are also aware of other organisations which are formulating plans for some form of coverage at 1:10,000 scale. In addition, organisations with rural interests are heavy users of 1:10,000 scale paper maps and some utilities have requirements for 1:10,000 scale map data where their plant is sparse, as in rural and semi-rural areas.

4.58 The market for digital topographic data is changing rapidly and market survey data is likely to become quickly out of date. We think it would be prudent if OS were to undertake further surveys to assess demand not only for 1:50,000 data but also for other derived scale digital data; it is over 4 years since the previous surveys were carried out and the OS needs more up-to-date information in marketing its products.

4.59 As with basic scales mapping, it is likely that individual users will go ahead with digitising the derived scales mapping to their own standards if some lead is not given by the OS. Again, we do not think that this will be in the long term interest of users because of the problems of duplication, incompatibilities of definition and the problems this will cause for exchange and sharing of data.

4.60   With the addition of digital contours, which are not included in the 1:1250 and 1:2500 basic scales data, the ideal solution would undoubtedly be to obtain digital data, equivalent to that held on the derived scale maps, by automated selection and generalisation from the basic scales data. This is particularly the case for 1:10,000 scale data since many of the 1:10,000 scale maps are now seriously out of date. Such automated selection and generalisation may be possible but will almost certainly require a much more sophisticated database and more intelligent selection and generalisation software than exists at present (see para. 9.14). The immediate need is for OS to make an assessment of the prospects for developing such software, as it is important if at all possible to avoid the problems and costs that separate digitisation of the 1:10,000 scale maps would create. Since the 1:25,000 scale map is produced from the 1:10,000 scale map, this should also be covered in the assessment. If this assessment shows that there is a reasonable chance that such software could be developed within, say, 5 years then OS should start a research and development programme to achieve this. Such a programme might be supported under the proposals made by the IT'86 Committee or through other Government support schemes for research and development (paras. 9.19-9.21).

4.61   There is, however, little prospect − even in the medium term − of being able to produce 1:50,000 type digital data by automated selection from basic scales digital data. We therefore accept that the only reasonable course, in the meantime, would be for OS to produce a separate digital data base from the 1:50,000 scale maps if there is sufficient demand for it. Before proceeding with the digitising, however, we believe it to be wise for OS to update their market surveys (4.58) and this must build on any feedback from the test of the second trial sheet which has now been digitised (para. 4.55). It is essential that the updating survey is based on affordable selling prices for the data; OS selling prices in the previous surveys were considered to be unacceptably high. It is possible that some of the substantial reductions in cost claimed for automated methods of digitising basic scales maps (para. 4.19) could also apply to digitising the 1:50,000 scale maps; moreover, we believe it to be unreasonable to charge the full cost of data conversion solely to the first user (see para. 4.62). The survey should include a consideration of the market for separate elements of the digital data and 'spin offs' such as digital terrain models; it should be carried out with urgency. Assuming a positive response to the new market survey, OS should define and publicise plans for the digitising and marketing of the 1:50,000 scale maps to create a small scales database; and should then proceed with the construction of that database. Wherever possible it should take advantage of existing digital data such as the contours digitised for the Ministry of Defence.

4.62   There remains the question of future pricing policy for derived scale digital mapping. It seems to us that there are good reasons, on both commercial and equity grounds, for not seeking full cost recovery in the short term. On commercial grounds alone, it seems realistic to plan for at least a 10 year payback period, and on equity grounds it would seem reasonable that the one-off initial conversion cost should be borne by users over a 10 year period rather than just by the initial users.

4.63   *We accordingly recommend that:*

1   *The OS longer-term objective should continue to be to achieve a scale-free digital database.*

2   *OS should conduct new market surveys of the future demand for derived scales digital map data.*

3   OS should make an urgent assessment of the prospects for developing the software to produce 1:10,000 type data from the basic scales database.

4   If the results of this assessment and the market surveys are positive, OS should support a research and development programme to develop the software to produce 1:10,000 type data from the basic scales data base.

5   Depending on the results of the market survey, OS should prepare and implement plans for conversion of the 1:50,000 map data to digital form.

6   OS should not be expected to recover the full costs of providing digital data at derived scales in less than a 10 year period.

7   To the extent that extra funds are required in the short term to implement these recommendations, this should not be at the expense of the basic scales conversion programme.

Concluding remarks   4.64   Ordnance Survey, as a Government Department, is subject to restrictions on its staffing numbers and gross running costs as are other Government Departments. We comment in the next Chapter on the adverse effect that this can have on the availability of statistical data. Such restrictions seem even more inappropriate to an organisation such as OS which has to operate in a commercial environment. When these restrictions are coupled with the financial recovery targets set for OS, they severely restrict the ability of the organisation to respond to the rapidly changing market situation.

We recommend that the current restrictions on OS staff numbers and gross running costs should be lifted: a restriction on net costs and a recovery rate target are all that are required.

4.65   When we started our Enquiry we found that the potential users of digital mapping felt that their requirements were unlikely to be met by the OS programme of digital mapping. We have been encouraged during the course of our work to find that OS has increasingly been prepared to adjust its policies and practices to meet the requirements of its customers. We mention three cases as examples which illustrate what we believe is a more responsive approach.

1   OS has recently concluded an agreement with Wessex Water Authority whereby the Authority will buy digital tapes from OS at a fixed price and then be able to sell them on to other utilities and users at the prevailing price for OS tapes; the Authority gets the tapes it needs quickly and OS gets a guaranteed income earlier than it would normally.

2   OS has recently concluded an agreement for Pinpoint Analysis to digitise and market Grid references of individual postal addresses from OS large scale maps; OS should thereby gain additional revenue from the creation of a new digital product from its existing paper maps.

3   OS is now conducting trials of alternative specifications and digitising systems as a matter of urgency.

We were also pleased to learn of new initiatives being taken by OS in the Geographic Information Systems and mapping fields as part of its continuing programme of research and development. These include liaison with academic institutes on such subjects as automatic feature recognition, urban information systems and digital terrain modelling. There is also continued growth of OS collaboration with the private sector in the development of in-car navigation systems, the testing of new equipment, the application of video-disc techniques and the use of the OS 1:625,000 digital database in commercial publications.

4.66   We are firmly of the view that active marketing of digital products will be all important in the development of digital mapping applications. It is essential that OS devotes adequate resources to this task.

5   Availability of data

5.1   Two main problems experienced by users trying to obtain or use spatial data — other than Ordnance Survey (OS) digital data — are:

1   obtaining spatial data in a form which permits linking to other data sets; this problem is most evident with Government data suppliers; and

2   trying to locate and access data from a variety of different sources.

5.2   To some extent the first problem arises from the policies and legal and financial considerations governing the operation of Government Departments. These affect many users of Government data and particularly users of spatial data. We consider these 'institutional constraints' together with the more general problem of locating and accessing different data sources, in this Chapter. Another cause of the first problem is the lack of standard spatial referencing of data made available, lack of standard spatial units for holding and releasing data and inadequate documentation of data. We consider these factors in Chapter 6.

**Difficulties in obtaining
Government data**

Constraints to releasing
data

5.3   Many of the Government Departments' data are not released for outside use or are not released at a fine enough level of geographical detail to meet users' needs. One of the main reasons for this is that there is often an extra cost involved in providing the data to 'outsiders'. For example, the data may be requested for smaller areas than those normally used by the Departments themselves and so extra processing may be needed.

5.4   A further serious constraint to releasing data, particularly unaggregated data, is the need to ensure protection of personal and commercial confidentiality and of civil and military security. Confidentiality of Government data is often protected by statute. Personal data in computer form, as elsewhere, comes within the scope of the Data Protection Act.

5.5   Another constraint to releasing Government data is the need to protect copyright. The main copyright issue expressed in the evidence was in relation to the OS and has been considered in Chapter 4. Of more general concern are questions of what constitute copyright of digital data and what obligations Government Departments have to protect Crown copyright and that of suppliers of data to them.

5.6   Witnesses referred to severe difficulties caused when data are no longer collected. For example, a number mentioned the withdrawal of the Census of Distribution which had caused them difficulty in planning or assessing the effect of new shopping developments. Some referred to the need for a mid-term Census of Population to provide reliable statistics on the population and characteristics of small areas. Although strictly outside our terms of reference, we draw these points to the attention of those concerned.

General approach to
supplying Government
data

5.7 Tackling the problems identified above depends to a great extent on the general policy adopted by Government Departments towards supplying information to outsiders. In recent years, following the recommendations contained in the 1981 report on *Government Statistical Services* (Ref. 14), Government Departments have concentrated on meeting their own requirements for data at minimum cost. In addition, Treasury rules have given little scope or incentive for meeting the demands of other users. We believe that these policies have inhibited the consideration of others' requirements for data at both the data collection and data handling stages.

5.8 There are signs, however, of a growing awareness within Government Departments of the commerical value of information. This change has been aided considerably by the setting up of a Tradeable Information Initiative, led by the Department of Trade & Industry, to encourage the release of more Government information of value to the private sector. Although at an early stage, the initiative has been active in encouraging Departments to take stock of their data holdings and others' requirements for them. Following inter-departmental consultation it has also provided Departments with a series of guidelines to assist them in dealing with companies wishing to acquire information (Ref. 15). Cost, confidentiality and copyright considerations are covered in the guidelines.

5.9 We welcome the Government's Tradeable Information Initiative and support the continuation of its work. We consider that its activities are pertinent to the release of Government data not simply to the private sector but also to the rest of the public sector.

Removing the obstacles
to supplying Government
data

5.10 The Tradeable Information Initiative has made a valuable start in drawing attention to factors constraining the release of data but further changes both in general approach and on some specific issues will be needed to achieve significant improvement in the supply of spatial data.

Basic principles

5.11 Considering the large sums of money spent on collecting spatial data held by Government Departments and the frequent lack of other sources of such data, we believe that Government held spatial data should be made as widely available as possible to maximise its use.

To achieve this we recommend that:

1 *Unaggregated spatial data held by Government Departments should be made available to other users provided that the costs of doing so are borne by the users and that there are no overriding security, privacy or commercial considerations.*

2 *If unaggregated data cannot be released because of the need to protect confidentiality, minimally aggregated data should be provided on a standard basis.*

Costs and charging

5.12 In considering whether or not to release information to outside users, each Government Department has to take account of what effect the extra cost of providing the data will have on its resources and whether or not it will receive any income from the activity.

5.13 For a number of Government Departments, a major obstacle to releasing data are the current Treasury restrictions requiring such activity to be contained within the ceilings on staff and gross running costs for each Department. In general, Departments are also required to pass any unanticipated receipts to the Treasury. These rules mean that, even when the extra costs incurred in supplying information would be covered by the additional income generated, Departments are often unable to meet the demand; even if they could, they have little incentive to do so.

5.14 Treasury and Central Statistical Office (CSO) guidelines require Government Departments to charge what the market will bear for data of commercial value. At the same time, costs can be waived in certain circumstances — for example — to local government when it has provided the original data. Although the Tradeable Information Guidelines reaffirm the principle of charging what the market will bear they also advise Departments that, for two years from the date of the Guidelines (August 1986), they may charge on a marginal cost basis — to recover only the extra costs of provision — where they are releasing information which has not previously been exploited.

5.15 *We recommend that:*

1 *Treasury restrictions on gross running costs should be lifted where Departments can show a net return.*

2 *Charges for data should be at marginal cost, and only at a higher rate if the market will bear it. For some users — such as educational establishments — reduced charges will be appropriate, as will be quid pro quo arrangements in some circumstances.*

5.16 One way of minimising the gross cost to Government of making data available is to let the private sector finance and carry out as much of the work of extracting and processing the data as possible. This can be done by franchising arrangements, subject to appropriate confidentiality restraints and quality controls. Organisations such as the Economic & Social Research Council (ESRC) Data Archive can also act as effective distributors on the Government's behalf.
We recommend that, subject to confidentiality considerations, there should be increased use of franchising and other arrangements by which outside bodies can act as distributors of Government data.

5.17 In some cases action taken by one Government Department to improve the provision of data will benefit other Departments requiring that data: the costs of improvements are not always shared in the same way as the benefits. Government accounting and administrative conventions should not be allowed to hinder the introduction of such improvements where the costs are borne by one Department and the benefits gained by several.

Confidentiality 5.18 In relation to all Government data, the Tradeable Information Initiative is encouraging Departments not only to observe normal Government rules on data security or privacy classifications but also to seek to avoid unnecessary obstacles to release of data. For example, it may be possible to use techniques which "disguise" the sensitive parts of such data, or to extract the sensitive parts where doing so will not significantly affect the usefulness of the data left. We also emphasise the importance of such measures, as an aid to achieving the release of minimally aggregated data. Where sensitive socio-economic data cannot be protected in this way the data should be aggregated and released on the standard basis we recommend in Chapter 6.

5.19 A number of statutes and other guidelines governing the collection and release of land ownership and land transaction data (described in Appendix 5),

Census of Employment and Census of Agriculture data were identified in evidence to us as being unduly restrictive.

5.20   Access to information held by HM Land Registry (HMLR) − on ownership, leaseholding, and related charges − is restricted to the registered proprietor or persons holding the proprietor's written authorisation. An open register would have a wide number of users, such as the utilities in extending or repairing their networks and property developers wishing to acquire land or property. The Law Commission of England and Wales in it second report on Land Registration 1985, (Ref. 16), recommended that the right of access to the register should be extended to everyone − as is already the case in Scotland and Northern Ireland and in almost all other countries. However, legislation to enact this has yet to be introduced.

5.21   There is substantial demand for information on individual transactions in land and buildings, including the price or rent paid. This information would aid rent reviews and company asset valuations, as well as the transfer of interests in property. Under present legislation covering England and Wales, the sale price submitted on Particulars Delivered Forms to the Inland Revenue cannot be divulged to a third party without the consent of one of the parties to the transaction. This is also the situation in Northern Ireland. This transaction information is, however, available to the public in Scotland and in many other countries. We understand from the Valuation Office (VO) of the Inland Revenue that there may be some scope for amendments to the present rules governing disclosure of aggregated information on property transactions, to reduce the level of aggregation used. It is our view, however, that releasing transaction data in any aggregated form would continue to significantly reduce the usefulness of the data to others.

5.22   In view of the usefulness of HMLR ownership data and VO transaction data for individual properties we believe there is a strong case for removing the constraints on releasing these data to any interested parties.

We recommend that there should be open access to details of land ownership contained in the HMLR's Register of Title and to details of land and property transactions held by the VO and the Valuation & Lands Office of Northern Ireland. This would bring England & Wales and Northern Ireland into line with Scotland. The necessary legislation to lift current restrictions should be introduced as soon as practicable.

5.23   Employment data are collected from firms on the basis of Government undertakings to protect the confidentiality of the information given and to use it only for the purpose for which it was collected. In the case of the Census of Employment, the restrictions are those embodied in the Statistics of Trade Act 1947 under which the Census is conducted. Although wider access − for example by local authorities − was provided under the Employment & Training Act 1973, individual responses to the Census cannot be made more widely or commercially available.

5.24   The supply of Annual Census of Agriculture data to outside users is constrained by similar considerations. In England and Wales, farmers and growers are required to supply this data under the provisions of the Agricultural Statistics Act 1979 (as amended by the Agriculture (Amendment) Act 1984), by completing postal questionaires. There are similar arrangements for the collection of the Census in Scotland. The accuracy and high percentage of returns achieved depend on the promise of confidentiality. In the longer term, remotely sensed data may provide an alternative source for some of the data now collected from farmers.

5.25   The provision of data from the Censuses of Agriculture and Employment is constrained not simply by the terms of the legislation referred to but also by the way in which the data are collected, being dependent on the goodwill of the data providers. Achieving less aggregated data in these two areas are likely to depend on finding alternative means of collecting the same or equivalent data, rather than on changing the legislation relating to current data collection methods.

Copyright

5.26   We consider that provision for the protection of digital data will be adequately covered when the 1985 Copyright Amendment (Computer Software) Act is reinforced by the legislation proposed in the Government's 1985 White Paper on *Intellectual Property and Innovation* (Ref. 17). This will also contain provisions to improve terms of Crown copyright and that of the original provider of information to Government.

We recommend that legislation for the protection of digital data, as proposed recently by the Government, should be introduced as soon as possible.

Assessing demand for data

5.27   Without some effort, it is often difficult for Government Departments to identify outside users' requirements for individual data sets. This is particularly so with regard to geographic information because of the immaturity of the market, the diversity of actual and potential applications, and the need for some users to link an individual data set to others before it acquires any value to them.

5.28   One-off studies of demand are usually carried out by or on behalf of Government Departments when changes affecting data provision are proposed. In some circumstances, there may be a need for wide reaching studies covering Government Departments, other public sector organisations, and the private sector. It is particularly important that outside demand for data is considered when computerisation of manual processes is proposed since computerisation may reduce the extra cost of meeting such demand. For example, computerisation of the VO's data on floorspace is likely to reduce the cost of providing data, suitably aggregated to preserve confidentiality, for outside users.

5.29   Assessment of demand can also be facilitated through the publication of lists of data held by Government Departments, and through individual Departments reviewing prospects for marketing each one of their data holdings. Both of these measures are being encouraged by the Tradeable Information Initiative. A more comprehensive and permanent measure would be the setting up of data registers as we consider later (para. 5.35). Assessing the market for linked data sets may require more active demonstrations of the possible 'added-value' achieved − for example by showing demonstration videos and computer programs to potential groups of users.

5.30   In some circumstances or areas, there may be a need for regular contact between Government Departments and existing or potential user groups. An example is provided by the Government's Information and Development Liaison Group which provides a forum for local government users of Government information. In Chapter 10 we consider the need for a further liaison channel, in the form of a central body, to complement these existing arrangements.

5.31 *We recommend that Government Departments should assess outside demand for their spatial data, particularly when computerisation of manual processes is to be introduced.*

Obtaining other public data

5.32 We believe that the aim of maximising the use of spatial data applies not solely to Government Departments but to all holders of spatial data collected at public expense.

We therefore recommend that other public sector organisations should adopt the basic principles to increase availability of spatial data which we have recommended that Government Departments should adopt. These are:

1 *that unaggregated spatial data should be made available to other users subject to cost and confidentiality considerations (para. 5.11);*

2 *that if unaggregated data cannot be released, minimally aggregated data should be provided (para. 5.11); and*

3 *that measures should be taken to assess demand for data (para. 5.31).*

Locating and accessing data

5.33 Much evidence was put to us on the difficulty users have in finding out whether spatial data on a particular topic exist, who holds the data and where, and how the data can be accessed. Related to these questions are those of how and which spatial data will be kept in the future, whether in the form of one central database or as a series of distributed databases; whether national or regional databases are required for particular sorts of spatial data or geographical areas; and whether data will be kept permanently in digitial form.

Distributed databases and data registers

5.34 It is clear that, in the future, digital spatial data will be held in a series of databases, maintained by national and local bodies — such as the national Land Registries and local authorities — and connected together by a nationwide communications network: this pattern is already evolving. The alternative of one central database for all spatial data would be impractical because of the large volumes of data it would be required to hold. It would also be undesirable, since it would be likely to limit accessibility and provision of advice on interpretation of data. Of fundamental importance to a series of distributed databases will be the setting up of data registers to help users locate and access data, plus the adequate documentation of data held. We consider this latter requirement further in para. 6.36.

5.35 There was significant support for a central register of geographic information or for registers for specific topics, such as data relating to environmental protection. The Royal Society also emphasised the importance of on-line access to such registers. At their simplest such spatial data registers could provide existing and potential users with knowledge of what data sets exist within particular databases, their content, and how they may be obtained or assessed. Even performing these functions alone such registers may need to be complex to be effective. Data registers could also help data suppliers assess the market for their data by providing information on demand for particular data, and could aid the development of standards for linking and accessing data. In the longer term they might be directly linked to the databases and provide an entry point for accessing data.

5.36 In view of the need to keep such registers up-to-date we believe that responsibility for them should lie with the data collectors or owners. We are aware that some data owners already have or are proposing to set up registers for data which they keep. We do not consider the creation and maintenance of one data register to cover all spatial data collected at public expense to be a practical proposition, although one register for all Government Departments might be possible and particularly valuable. We believe there is a need for a central point which could direct initial enquiries to individual registers. A suitable location for this might be the British Library which already holds a register of digital cartographic data sets.

5.37   *To help users locate and access data, we recommend that:*

1   *Holders of spatial data collected at public expense should set up data registers individually or in groups as appropriate.*

2   *The CSO should consider the setting up of one register for all Government Departments.*

3   *The CSO and the British Library should consider the setting up of a central direction point to individual registers.*

National and regional databases

5.38   In some cases one central national database to cover a particular area of interest may be desirable to avoid duplication in holding data, or to provide skilled staff to aid interpretation. For example, the ESRC is developing a national database of selected social sciences data. The form of such central databases — whether attempting to be comprehensive or holding key data sets, providing on-line or other access — should be determined largely by users' requirements.

5.39   The ESRC is also in the initial stages of setting up a rural database which might contain a variety of socio-economic, rural land use, and environmental data. We welcome this initiative, and we are convinced — because of its multi-disciplinary nature — that close collaboration between the various agencies who can supply the data and potential users of the database will be essential for its success. We look to the ESRC to ensure that such collaboration takes place.

5.40   We see some advantages in the setting up of regional centres which pull together national and local data sets for the area or region. One of the functions of the four Regional Research Laboratories, recently established with ESRC funding and based in higher education institutes, will be to hold selected spatial data sets for particular regions. They may provide a useful indication of demand for such centres.

Permanent information

5.41   A great deal of geographic information, as with other information, is required permanently or for long periods of time so that it can be used, for example, for historical or 'time series' studies. Permanent or archival data has traditionally been kept in a variety of paper forms — print, manuscript or maps — by bodies such as the British Library, the Public Records Offices and Companies House. The archiving of digital data by such bodies is not well established and faces a number of problems:

1   Under existing copyright legislation the publisher of every book published in the United Kingdom must deposit a copy with the British Library and with any of five other specified libraries if they call for a copy. There is currently no such legal requirement to deposit copies of digital data. The Government's White Paper on *Intellectual Property & Innovation* (Ref. 17) concluded that legal deposit needed to be reviewed in a different context to copyright. To allow this to be done the Office of

Arts & Libraries assumed responsiblity for legal deposit in April 1986 and will consider the scope and timing of such a review. In the absence of any legal requirements to deposit digital data, libraries and other archiving bodies may require significant funds to purchase the data.

2   In addition, the development of digital storage facilities will require considerable investment.

3   A further problem is that of selecting digital data for archiving: with increasing amounts of digital data being generated there is a need to be selective in the data archived.

5.42   Following the recommendations of the House of Lords Select Committee on Science & Technology and the Government's response (Refs. 1 and 2), the British Library and the other legal deposit libraries were made responsible for establishing an archive of digital mapping data and a reserve archive of remote sensing images, the main archive of remote sensing data being held at the National Remote Sensing Centre. The deposit libraries and other archiving interests are currently considering the problems involved in establishing these archives.

5.43   To safeguard the retention of other key spatial data sets in digital form, we believe that users, suppliers and holders of particular sorts of spatial data − such as geological or socio-economic data − should agree which data in their area of interest should be kept permanently and by whom. The central group we propose in Chapter 10 could monitor progress in this area.

6 Linking data sets

6.1 An important feature of applying Information Technology to the handling of spatial data is the ability to link data sets; that is, to merge and compare different data for the same location. This makes a Geographic Information System an analytical and decision making tool fundamentally different from a paper map. It is the ability to manipulate readily and quickly large volumes of data for the same areas which has the potential for adding great value to spatial data.

6.2 The importance of linking data was recognised by a wide range of users who submitted evidence. The many applications described in Chapter 2 depend on such data linkage. Several of these applications require linking of data across traditional areas. For example, the recent growth of epidemiology has required the integration of socio-economic and environmental data. To realise the full utility of remotely sensed data, particularly for analysing environmental characteristics, it is vital that they are linked with data from other sources.

6.3 At present, it is often impossible or very difficult to link data sets and consequently to draw statistical inferences from them, as illustrated in *Box 6A* overleaf. These difficulties can arise because:

1. the data are not explicitly spatially referenced, the data relate to a variety of different areas which do not match or nest into each other and the boundaries of these areas are frequently changed for administrative reasons. These factors are considered in greater detail in Appendix 7:

2. data sets are poorly or inconsistently documented; and

3. different users' computer systems — including hardware, software, and data structures — and communications equipment are incompatible.

6.4 In this Chapter, we examine how these difficulties can largely be overcome by adhering to common standards. We emphasise the importance of Government, as the major provider of data, adopting these common standards.

Standards for locational referencing and spatial units

Locational referencing

6.5 Geographic information can be represented by three basic features — points, lines or areas. For example, a house might be represented by a point, a road by a line, and a forest by an area. Each feature needs to be locationally referenced to identify where it is on the land, in the sea, and sometimes also above or below them.

6.6   Almost all the evidence to us on this aspect recommended the referencing of land-based information to the National Grid or Irish Grid and marine-based information to latitude and longitude, as the most flexible and commonly used means of referencing data for storage and subsequent use. Such referencing may be direct through the inclusion of Grid references or indirect through the use of cross-reference tables which link the locational referencing system used to the National Grid, devised by the Ordnance Survey (OS), or the Irish Grid.

6.7   For the land area of Great Britain, the National Grid provides a way of allocating locational references for any point lying within its framework. These references take the form of numerical coordinates for eastings and for northings: the number of digits used varies according to the level of precision required. Lines and areas can also be locationally referenced in this way: for a line by a series of point references, and for an area by using the coordinate references for the centre of the area or points describing its boundary line. In addition, vertical references or heights can be provided. For Northern Ireland, features can be locationally referenced in a similar way using the Irish Grid, which can be computationally correlated with the National Grid.

6.8   Marine-based data can be related to the traditional referencing system of latitude and longitude and can additionally be tagged by depth measurements. Using a simple conversion formula, National and Irish Grid coordinate references can be converted to latitude and longitude and vice versa.

6.9   *We recommend that, as far as practicable, all geographic information, including remotely sensed data, relating to the land areas of the United Kingdom should be referenced directly or indirectly to the National Grid or Irish Grid as appropriate.*

Spatial units   6.10   Difficulties arising from different and changing geographic areas, or spatial units, for which data are released were extensively referred to in evidence to us. At present, a large number of spatial units are used to store or present data. Many spatial units of similar size have different boundaries and therefore cannot easily be matched and many do not nest into larger areas. This makes aggregation of data and comparison of areas difficult. In addition, administrative boundaries are often changed making it hard to detect trends over time. These problems occur especially in relation to socio-economic data where there is often a need to link data relating to individuals, households and groups of people living in a particular area.

6.11   If data were held or released in unaggregated form, it would be relatively easy for data suppliers or users to aggregate the data to the geographic areas they require for their own purposes.

We therefore recommend that data suppliers should both keep and release their data in as unaggregated a form as possible.

6.12   Where this principle cannot be followed, as is often the case for confidentiality reasons which we considered in Chapter 5, we believe the practical solution would be for data suppliers to hold or release that data on the basis of addresses and unit post codes. These would give the flexibility necessary for aggregation into larger geographic areas of different shapes and sizes required for particular purposes. We outline our two preferred bases below.

Box. 6A: Difficulties in linking and analysing geographic information

Linking data

The need to link data sets together is a common one and stems either from the use of data for purposes other than which they were originally collected or, just as frequently, to save costs of data collection by exploiting what already exists.

Spatial data are linked together through a knowledge of their position in space. This can be simple — two pieces of information related to the county of Durham, for instance, are easily linked together — provided they relate to an unchanged area, such as most counties after 1974. Equally, it is easy to link data for areas which form part of a spatial hierarchy e.g. where some information relates to electoral wards and some to administrative districts (since the former fit or 'nest' within the latter). The greatest difficulty in data linkage occurs where there is no match between the spatial units of two data sets to be combined: this is illustrated by the following example concerning environmental data.

If a combination of soil and geology is to be carried out so as to examine the effects of the underlying materials on the surface soil, then some approximation is required. In essence, some assumption has to be made — such as that the soils and geology are uniform within each mapped area; the data are then disaggregated into areas with one type of soil and of geology.

Overlay

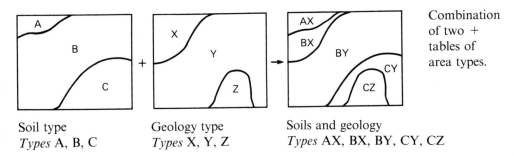

| Soil type | Geology type | Soils and geology | Combination of two + tables of area types. |
| *Types* A, B, C | *Types* X, Y, Z | *Types* AX, BX, BY, CY, CZ | |

There are technical difficulties which constrain the effective linkage of spatial data. Sometimes, for instance, the description of an area is incomplete: the 1981 Census Enumeration Districts were described in computer form only by the approximate centre point National Grid reference of each of these areas, which vary greatly in size. In other cases, the areas may be fuzzy in having no unique definition (such as 'the Pennines'). Of particular difficulty when dealing with parliamentary, economic or manpower data is that the basic 'building blocks' for which data are available mutate subtly over time.

Analysing data

Though they share many of the characteristics of other data sets which are statistically tabulated and analysed, spatial data have additional characteristics which can cause major problems in analyses, particularly those which attempt to draw statistical inferences from linked data sets. These problems include:

1 *The modifiable areal unit problem:* The results of applying *any* statistical method to geographical data are influenced to some, usually unknown, degree by the nature and number of the areal units — such as counties or grid squares — under study. In a statistical sense, these units are arbitrary: in reality, rarely do distinct breaks in characteristics such as population density and land use coincide with the edges of the areal units for which the data are collected. The relationship between such characteristics can vary dramatically if the boundaries of the areal units are changed. There are two different components to this problem. First, the many different ways of grouping the same small areas — e.g. wards — into larger ones — e.g. parliamentary constituencies — can all give different results. Second, analysis at different levels of resolution — e.g. counties as opposed to districts — will often tend to give very different results, notably for the relationships between variables. Relatively little work has been carried out on this problem: most studies have been empirical. Hence, there is little confidence about the results of many analyses unless they are confirmed through many repetitions using modified areal units.

2 *Spatial auto-correlation:* When neighbouring data values are highly correlated — e.g. adjacent areas have similarly high or low unemployment rates — the data are said to possess high spatial auto-correlation. This property clashes with the assumption of independence underlying many statistical procedures, especially interval estimation and those hypothesis tests which produce attained significance levels. Results from these methods are likely to be distorted to some unknown degree by the presence of spatial auto-correlation. There is no simple remedy for this other than to avoid completely the use of inferential statistics. There has been some limited success, over the last 20 years, in devising more appropriate methods for statistical analysis of spatial data. But much additional research in this area is required to provide easy-to-use and reliable tools.

3 *Spatial sampling:* In recent years, much use has been made of sampling to provide low-cost national estimates — e.g. in the DOE Landscape Change project total hedgerow removal was estimated from sample points. Two problems arise here concerning the distribution and number of sampled points. First, the geographical distribution of the sample points can be chosen on many statistical criteria: in the real world, however, physical access to the possible sample points varies so greatly that we need measures of the trade-offs involved — e.g. how much statistical reliability do we lose if we choose convenient rather than ideal sites? Second, to produce reasonable national figures may require only a sample of about 1000 households but to produce the same statistics for districts might require the same size sample in each of the 458 districts.

The address 6.13   The concept of a standard spatial unit for holding and releasing data was last examined by the Government and local authorities in the *General Information System for Planning* (GISP) initiatives in the early 1970's (Ref. 18). The 'Basic Spatial Unit' then recommended was the rating hereditament for urban areas and OS land parcels for rural areas.

6.14   We agree with the need for a standard spatial unit for linking socio-economic data at about the level of the individual property. However, we prefer the use of the postal address although without defined boundaries it is not strictly speaking a spatial unit. For most socio-economic data this presents no problem as boundary information is not relevant. The advantages of the address are that it relates better to individual and household data in multi-occupied properties and that it ties in with the post code system which we recommend as the next level of standard spatial unit.

Unit post codes 6.15   We also believe that a larger unit is required for data which cannot be released at the address level because of the need to protect confidentiality, and to avoid the administrative difficulties of maintaining large databases and updating data at the individual address level. We believe the most appropriate area at this level to be represented by the unit post code, although again this is not truly a spatial unit unless its boundaries have been defined.

6.16   The unit post code, comprising a string of addresses, is the smallest of a hierarchy of 'areas' used by the Post Office for sorting and delivering mail. The boundaries of unit post codes are not yet defined in England and Wales although they are mapped manually in Scotland. There are 1.5 million unit post codes in the United Kingdom with an average of 14 addresses in each unit post code. Appendix 8 outlines the post code system, including the Post Code Address File (PAF) maintained by the Post Office, and the Central Postcode Directory (CPD) maintained by the Census Offices for England and Wales, and Scotland. The PAF and the CPD permit data to be cross-referenced to other spatial data.

6.17   In the submissions of evidence to us, there was considerable support for wider use of the unit postcode as a spatial unit for address related statistical data. Many Government data sets are now referenced by unit post code; the Central Statistical Office considers the post code system to be the preferred method of allocating Government socio-economic data to small areas. The 1981 Census of Population for Scotland was organised by the General Register Office, Scotland (GRO(S)) on such a basis, and other statistics released by the GRO(S) have been referenced by unit post code for several years. As we described earlier (2.26), this has facilitated the linking of medical and population statistics in Scotland. In the private sector, unit post codes are already used in a variety of applications.

6.18   We believe that the unit post code, used within the post code system, has many advantages. It is sufficiently small to facilitate aggregations to larger areas and to locally defined areas of interest; it permits the linking of address related data whilst preserving confidentiality; and its use by the Post Office as an established working system ensures that it is kept up-to-date and provides full coverage of the whole of the country.

6.19   There are a number of practical problems associated with the use of unit post codes as standard spatial units. Addresses covered by a unit post code do not always nest precisely into administrative and ward boundaries; many changes are made to unit post codes (approximately 18,000 per year) by the

Post Office; and local Post Office administrative variations sometimes result in inconsistencies in defining unit post codes. In addition, the dispersed nature of the rural population results in unit post codes which cover very large areas yet may need to be merged with others when presenting socio-economic statistics to protect individual confidentiality.

6.20   It might be possible to reduce slightly the first problem of mismatch if unit post codes were to be taken into account by those setting administrative and ward boundaries. The setting of ward boundaries so as not to split unit post codes would require legislative change for which we do not consider there is a strong enough case. However, we urge the Boundary Commissions to take account of unit post codes when setting ward boundaries as far as is consistent with their statutory obligations. Conversely, we would ask the Post Office when establishing new unit post codes to take account of topographic features and electoral and administrative boundaries. Other problems may be reduced by improvements being made by the Post Office to the operation of the post code system − we further consider its role in this respect in para. 6.34 − and as a result of the proposal to Grid reference individual properties (referred to in para. 4.65).

6.21   The address can be linked to the unit post code, and vice versa via the PAF. There was some support in the evidence for an additional reference for properties, i.e. a unique property reference number (UPRN). Many existing property information systems include a UPRN (leading examples are shown in *Box 6B*). But we can see no advantage in having a national UPRN system in addition to an address for each property.

6.22   *We recommend that the preferred bases for holding and/or releasing socio-economic data should be addresses and unit post codes. Wherever possible, the boundaries of administrative and electoral areas should not split whole unit post codes.*

Grid referencing addresses and unit post codes

6.23   To locate the two recommended standard spatial units and permit linking or nesting of their data with those of other administrative units it will be important to attach a Grid reference to each. Both addresses and unit post codes could be Grid referenced by a single point or by delineating and then digitising their boundaries.

6.24   Although a number of local authorities maintain computer-based land and property databases which contain point Grid references of properties − for example the Joint Information Service of Tyne and Wear − point Grid referencing is not generally available. However, it is likely that point Grid references for individual addresses will be available commercially in the next few years through the project, referred to previously (para. 4.65), which involves the OS. For certain activities, such as the recording of land ownership and development planning, precisely delineated boundaries of the property are required. A small number of local authorities − such as South Oxfordshire District Council and Dudley Metropolitan Borough Council − maintain digital records of property boundaries and the HM Land Registry is considering whether to digitise its own land title boundaries, currently defined on OS paper maps.

6.25   Point Grid referencing of unit post codes is currently available through the CPD. The nominal levels of resolution are 100 metres in England and Wales and 10 metres in Scotland, but we have been told that there are many inaccuracies. It should be possible to improve the resolution of unit post code

Box 6B: Property referencing systems — some examples

Local Authority Systems: In the 1970's a number of local authorities established several types of property information systems. This was encouraged by Government, for example through the publication of the report on Geographical Information System for Planning (HMSO 1972) and by the DTI's support for the development of the Local Authority Management Information System (LAMIS) by ICL. A common feature of such systems was the use of a Unique Property Reference Number (UPRN) which acted as an address code and allowed the unique identification of every property within the area of coverage. Such UPRNs provided the common reference to link together different files containing information relevant to property and planning.

The LAMIS system contained UPRN's based on rating hereditaments in urban areas and on physical ground features in rural areas. The National Gazetteer Pilot Study, also set up in this period and eventually developed into the Joint Information System (JIS) of Tyne and Wear, included UPRN's based on properties or land parcels distinguished by full postal address.

National systems: On a national level, the Valuation Office prepared a computer based scheme in the late 1970's to reference all rateable properties by allocating unique property reference numbers (VOPREN). This scheme was abandoned when the 1982 revaluation of properties was cancelled. However, the Valuation Office continues to maintain a manual index of such reference numbers.

The HMLR allocates unique reference numbers to each individual land title. The land in a title may consist of a single property, e.g. a house or farm, or of only a part of such a unit, or a number of such units. The HMLR's land title reference numbers would not therefore provide an adequate UPRN for many local authority applications.

The Post Office's PREMCODES: The Post Office aims to provide a unique property reference based on the unit post code and the PREMCODE, i.e. the first four alphanumeric characters of the address, typically the postal number of the premises and the first letters of the thoroughfare name. Where the premises have a name, the PREMCODE consists of the first four letters of the name. These rules usually make a PREMCODE unique but, where an ambiguous PREMCODE would occur, it is allocated a pseudo code to ensure uniqueness. The format of the Unique Property Reference is therefore typically an eleven character reference: AANN NAA NNAA. Where A = Alpha and N = Numeric characters.

point references if point references become available for all addresses. The Post Office is currently considering the demand for digitised unit post code boundaries. As boundaries have already been defined manually in Scotland, it might be advantageous to provide digitised boundaries for Scotland in the first instance, should sufficient demand for digitised boundaries become apparent.

6.26 *We recommend that all addresses and unit post codes should be Grid referenced. For addresses, the proposed provision of point references should meet most requirements. For unit post codes the current point references should be improved to meet users' requirements. To avoid different Grid references for the same address or post code, Government − probably the OS and/or the Census Offices − in conjunction with the Post Office, should be responsible for ensuring consistency in Grid referencing of addresses and unit post codes.*

Spatial units for rural resources and environment data

6.27 The spatial units used in most traditional representations of environmental characteristics − such as soils, geology or vegetation cover − differ from those previously considered in that they are naturally defined. They are normally complex polygons whose boundaries are determined by the phenomenon or use being mapped and by the particular classification system being used. These 'natural' units of different characteristics often do not coincide, as is illustrated in *Box 6A*. In some cases the data collected will relate more usefully to management areas such as farms or forestry compartments: here National Grid referencing will be important.

6.28 Often it is necessary to make areal estimates using limited field observations. In such cases a sampling scheme based on grid squares has been found useful by the Institute of Terrestrial Ecology for making estimates − on a national and regional basis − of phenomena as diverse as changes in the length of hedgerows and the impact of fall-out arising from the recent accident at Chernobyl nuclear power station. Regular grids are also of increasing relevance with the development of remotely sensed data where the smallest unit of collection − as small as 10m from satellite sensors − is the square pixel.

6.29 Sampling schemes are often desirable on cost grounds when collecting rural resources and natural environment data for national or regional aggregates. Where natural spatial units for different phenomena do not correspond care must be taken in defining relationships.

Promoting standards for spatial units and locational referencing

The role of Government Departments

6.30 We consider that the most effective way of achieving more widespread use of our recommended standards for locational referencing and bases for holding and releasing data is through their application to Government data. In the area of socio-economic data the Census Offices for England and Wales, and Scotland have by far the greatest influence of any Government body in setting standards both outside and within Government. We therefore consider it to be of the utmost importance that Census of Population data are held and made available on a unit post code basis. Because of the need to preserve confidentiality, data − apart from population counts − could not generally be released for individual unit post codes but could be made available on an aggregated basis to meet particular requirements.

6.31 In 1985, the two Census Offices introduced proposals for a post code based geography in the 1991 Census, to provide a basis for collection of the Census and analysis of results. The proposals had two main phases: firstly, each postcode unit would be mapped as a zone on large scale OS maps, adopting the methods already established by the GRO(S) and secondly, the zone boundaries would be digitised and a digital database set up to permit more comprehensive manipulation of data, including mapping of results. Although the cost was higher than traditional methods of conducting the Census, there were believed to be additional benefits in the longer term. However, in view of pressure on public expenditure, it was decided not to proceed further with mapping post codes in England and Wales. The Office of Population Censuses & Surveys (OPCS) is now examining other ways of offering postcodes as a geographic base for small area statistics from the 1991 Census. The GRO(S) will continue to use post codes in pre-Census planning and as a base for output of Census statistics.

6.32   We were most disappointed that the proposal for a fully post code based 1991 Census in England and Wales was withdrawn, particularly since this action was taken without a full survey of potential users to clarify what direct or indirect short or longer term financial benefit could have been expected. We understand that there may not now be sufficient time before the 1991 Census to allow the reintroduction of the proposal for planning the geography of Enumeration Districts on a unit post code basis. However the analysis of the 1991 Census data on a unit post code basis could still be achieved provided the OPCS referenced the data for individual households by unit post code.

6.33   We consider that the way in which OPCS data is made available will be the main avenue for setting spatial referencing standards for socio-economic data.

We recommend that the OPCS, in addition to the GRO(S), should ensure that the results of the 1991 Census of Population and any future censuses are available, subject to confidentiality, on a unit post code basis.

The Post Office's role   6.34   The wider use of the post code as a national referencing system will place responsibility on the Post Office to maintain and develop the system to meet users requirements in both public and private sectors. As we identified earlier (para. 6.19), there are still a number of significant difficulties to be overcome before the unit post code can operate to best effect as a standard unit for holding data. The Post Office has regular discussions on the use of the post code system with Government Departments and local authorities and has recently commissioned a major study of public and private sector requirements in this area with a view to improving the system. However, in view of the importance of this system, we believe that there is a role for a central body to review and if necessary take action to help the Post Office's activities in this area. This role could be assumed by the body we propose in Chapter 10.

6.35   We welcome the Post Office's recent moves to improve the post code system. We consider that such improvements will benefit both public and private sectors. It will be essential for the Post Office to review users' requirements on a regular and comprehensive basis particularly when proposing changes to the post code system.

Standards for data documentation

Non-cartographic spatial data   6.36   The data sets which a user may wish to link are likely to vary greatly in quality, terms and definitions used. Without adequate documentation, the user cannot tell how serious are the differences between data sets. We consider that all data sets should be adequately documented to indicate factors such as the quality of the data — for example, date and method of collection, indication of reliability and definitions used. Specialised bodies such as the Economic & Social Research Council Data Archive, which is developing documentation standards for social sciences digital data, and the British Library are well placed to give advice in this area. We believe it is particularly important that standards for documentation of Government data sets are developed so as to set standards for other data suppliers to follow.

6.37 *We recommend that the Central Statistical Office and the Department of Trade & Industry should encourage and lead the development of data documentation standards within Government. Non-Government data suppliers should follow any Government lead in providing similar high standards of documentation for data sets.*

6.38 It is also difficult to link or compare spatial data sets if they are described by different classifications. For many purposes a number of classifications have evolved — for example, to describe land use or soil type — to suit the needs of particular users; in these circumstances some coordination would be desirable. We consider this to be the responsibility of those concerned with the collection and analysis of geographic information in particular areas such as land use, utility plant and equipment, agricultural land quality and road networks. We have been impressed by the activities of the National Joint Utilities Group in drawing together the common interests of the utilities in this respect. It is important that classifications are well documented.

Digital map data 6.39 The handling of digital map data introduces special problems because there is a fundamental distinction between the digital representation of cartographic data and its paper form. In the digital form, there is a requirement to encode information explicitly whereas in graphical form many details — such as where the edge of a river may coincide with a boundary — are implicit. A second distinction between the traditional map and its digital representation is the need to encode numerically the attributes of the cartographic data which are normally conveyed to the reader of the printed map by colour, line weight, symbols or labels. This can be done in a number of different ways; without governing standards a variety of incompatible schemes would undoubtedly arise.

6.40 In Great Britain, standards covering the transfer of digital map data are being developed by a National Working Party, set up in February 1985. It is led by the OS and comprises representatives of Government, the utilities, local authorities, research interests and professional institutions. Although the Working Party's main aim is to define national standards for the transfer of digital map data, its remit is wider than this since it was recognised that the definition of the data transfer format was of limited practical use without providing a definition of the subject of the transfer, i.e. the map data. Hence the Working Party has been studying data transfer formats, feature classification, data quality descriptions including lineage, and a glossary of terms.

6.41 The final draft standards were published in February 1987, and the OS has offered to provide resources to allow their adjustment to meet user requirements in the following 12 month period. Thereafter the British Standards Institution, which has the remit to promulgate and update national standards, may adopt them. Their widespread use by the Ordnance Survey would in any case help to establish them.

6.42   We welcome the speedy production of draft standards for the transfer of digital map data by the National Working Party. It is important that the draft standards are given wide circulation, with adequate time to allow views and comments from interested parties.

We recommend that maintenance of the adopted standards for the transfer of digital map data should not be a charge against OS resources, if OS undertakes this role.

6.43   We note that the National Working Party's draft standards can in principle be used for transforming remotely sensed data although in practice this can currently be cumbersome due to the large volume of remotely sensed data. There is a clear need for standardisation of remotely sensed data transfer in its various forms — such as tapes and optical discs. The National Remote Sensing Centre is developing transfer standards for such data. We ask it to ensure that these standards are accomplished and that they are compatible with the National Working Party's draft standards insofar as the rather different nature of remote sensing data allows.

International standards

6.44   International standards for the exchange of digital cartographic data are under development within the NATO context. We note the involvement of the Directorate of Military Survey, Ministry of Defence, in this work which should lead to the formulation of standards compatible, although not necessarily identical, with those being formulated by the National Working Party (para. 6.40). The NATO proposals for digital data transfer standards are also similar to those developed by the National Committee for Digital Cartographic Data Standards in the United States and are likely to have an important influence on the development of Geographic Information Systems.

6.45   In the case of hydrographic charts, the international framework is well established through the International Hydrographic Organisation and its various member states hydrographic offices. The exchange of marine information has a long tradition and the transition to digital data with the relevant exchange standards should present no insuperable problems.

6.46   Attempts have been made to standardise some classifications on an international basis, such as those defining land use categories. These attempts have, in general, not resulted in unique standards but have served to provide a general structure within which variations to meet specific purposes can be developed and the limitations on the techniques used to collect information can be taken into account.

6.47   We believe that there is a strong case for using an internationally approved classification, especially where international exchange or comparison of data is important. It would be helpful if any national classification scheme which is used could be aligned with the international standard.

6.48   We consider that the goal of agreed national and international standards for the exchange of digital spatial data and provision of full documentation of the various terms and categories which form the basis of such standards is of considerable importance and efforts to that end should be encouraged.

Information technology standards

6.49   Both the UK Government and the computer manufacturing sector support the development of common standards in the field of Information Technology, including the Open System Interconnection (OSI) standard. Such standards will be vital to an organisation handling geographic information where there are requirements to use different machines for handling the same information and to exchange information with other organisations. The OSI framework, when fully developed, should ensure that specialist equipment can be bought with some confidence, that the complete system will perform as required and that expensive equipment can be shared in a flexible manner. It should also help to avoid the situation where the user is forced to continue purchasing a particular manufacturer's products and services.

6.50 The Graphical Kernel System (GKS) is the computer graphics standard promoted and adopted by the British Standards Institution and provides the basic structure to be used in software for presenting information in graphical form. The aim is to have the system adopted by the producers of software so that machine-independent software can be developed and a wide range of graphics input and output devices can be used without modifying the users' software.

6.51 We note that work on the development of OSI and GKS standards is well advanced and such standards are gradually being adopted. We endorse the need for such standards in the area of geographic information handling and support their promulgation by the British Standards Institution and the International Standards Organisation.

7 Awareness

7.1 We believe that one of the most important reasons for the low take-up of Geographic Information Systems is a lack of awareness of the potential benefits to be gained from their application. We consider here possible reasons for this and why we believe there is a case for promoting awareness. We outline our views as to how this could be achieved through a series of measures which demonstrates the benefits and applications of Geographic Information Systems. A further important avenue for spreading awareness is through the education and training process. We examine this in Chapter 8.

Reasons for lack of awareness

7.2 We found evidence of a lack of user awareness at several levels. These range from ignorance of the techniques themselves and the benefits of using them to − at the other end of the scale − ignorance of how best to go about using these techniques. Since relatively few people in the United Kingdom currently use Geographic Information Systems there is little information on users' experiences of, for example, how well the systems performed certain tasks, what the benefits − expected and unexpected − were and what were the main obstacles to their introduction. In some areas, overseas experience might be valuable to potential UK users but is not generally known. A further problem is that the rapidly changing technology and systems available make it difficult for users to keep their knowledge up-to-date.

7.3 There is also a lack of knowledge of data sources and how they can be accessed. This is partly because many data suppliers have not publicised the data they hold, even within their own or related organisations. We considered measures which data suppliers could take to overcome this problem in Chapter 5, but note here that data suppliers will first need to appreciate the potential market for, and value of, their data when held in Geographic Information Systems.

The need to promote awareness

7.4 We believe that lack of awareness of the current state of the art in handling geographic information is not simply holding back its wider use. It can also result in duplication of investment and repetition of the learning process, both within and between organisations. The full benefits of sharing geographic information and Geographic Information Systems cannot be realised unless all the potential sharers are aware of them.

7.5 Without an appreciation of the possibilities, potential users are unable to specify properly their requirements and system suppliers are unable to identify users' needs clearly. The result is that system suppliers are slower in developing the right products. If requirements are not defined accurately, there is a danger of developing over-ambitious systems, developing systems to meet the wrong goals and purchasing the wrong sort of systems.

7.6 All this is part of the much wider lack of awareness of the potential and benefits of Information Technology. The recent report from the Advisory Council for Applied Research and Development (ACARD) *Software: A Vital*

Key to UK Competitiveness (Ref. 11), highlighted the generally slow take up and application of Information Technology innovations and considered general knowledge of Information Technology to be a key to rapid acceptance.

7.7 Measures to promote Information Technology, such as those recommended in the ACARD report and those already being undertaken by Government and the private sector, should help increase users' receptiveness to the new opportunities for handling spatial data. In addition, there is a need to promote the use of geographic information technology in its own right.

7.8 Suppliers of Geographic Information Systems have an obvious interest in promoting their use and no doubt they will pursue their traditional marketing and product development activities with vigour. However, in view of the diversity of interests in the geographic information field and the rapidly changing technology and systems available, we believe additional promotional measures should be taken.

Action to promote awareness

Overall strategy

7.9 We see two main avenues through which to promote the use of geographic information and Geographic Information Systems. The first is through the general education and training process, which should provide students with an initial appreciation of manipulating spatial data and modern techniques for doing so. The second is through a series of measures, which we outline later, based on demonstrating the uses and benefits of geographic information technology and quickening the flow of information on current systems and data sources. The main aims of these measures are to:

1 help users appreciate the applications and benefits of Geographic Information Systems; and

2 help users make the best use of Geographic Information Systems.

Priority user targets

7.10 The primary targets of these promotional measures will generally be users but data suppliers, researchers, trainers, and system suppliers will also benefit from them and may themselves be the main targets of particular measures.

7.11 There are a number of key groups or 'seeding points' who should be the focus of promotional efforts aimed at users. One group of significance in initiating the use of Information Technology comprises the heads of organisations — for example, Chief Executives in local government and Heads of Departments in Central Government. It will be important to gain the confidence and enthusiastic support of such people by demonstrating that the output from Geographic Information Systems can result in better decision making and resource management.

7.12 But it is also important that the promotional measures are aimed at:

1 other levels of management, such as the senior managers who might be responsible for providing the information demanded by the Chief Executive;

2 at politicians and trade unionists who also play key parts in the introduction of new systems; and

3   at educationalists who are in a position to influence the users of the future.

Implementation

7.13   Carrying out the promotional measures we recommend will involve system and data suppliers, as well as educationalists and researchers and users themselves. Communication between users, including through user clubs such as the National Joint Utilities Group and Local Authority Associations, provide an effective means of spreading awareness. In many cases, promotion of awareness will be best achieved by consortia of suppliers, users and others.

7.14   In view of the breadth of the field and number of parties involved, we believe there will be a need for a body to provide a central focal point for promotional measures. In addition, it could provide support for the main parties undertaking the promotional activities we recommend; users in particular may have limited time and money to promote their experiences. It could also investigate additional methods of promotion to those identified here. This role and related activities could be carried out by the central body we propose in Chapter 10.

We recommend that some central support should be established to provide a focus and encouragement for measures to promote the use and development of Geographic Information Systems. This should be in the form of the central body we propose in Chapter 10.

Specific Measures

7.15   We identified in Chapter 5 (para. 5.29) a number of measures which would promote knowledge of data sources, including the publication of lists of data, video and computer demonstrations, and the establishment of data registers and a central point of contact. Measures to promote awareness of the applications and benefits of Geographic Information Systems are considered below.

Dissemination of information

7.16   Important means of promotion, currently little developed for disseminating information on geographic information technology and its application, are articles in professional and other general interest journals, newsletters of user groups and societies, published reports and case studies, videos and computer based demonstration packages, seminars, and conferences. It is essential that information is disseminated via journals and conferences attracting the attention of the potential users and key decision makers. Such information would include:

1   information on current and proposed applications of Geographic Information Systems, including associated costs and benefits and evaluation methodology;

2   the results of research and development within the United Kingdom and its relevance to particular application areas; and

3   information on the best overseas work.

We recommend that user groups, data and system suppliers, and academics should actively disseminate information, via appropriate media, on the applications and benefits of Geographic Information Systems.

Familiarisation courses

7.17   One method of introducing potential users to the uses and benefits of geographic information technology is via short courses and workshops of 1-3 days in length. There is heavy demand for the few familiarisation courses currently run.

We recommend that more Geographic Information Systems familiarisation courses should be provided, perhaps on a joint funding basis by system and data suppliers and academic institutions.

Demonstration projects

7.18   There was much support in the evidence for projects to demonstrate various aspects of handling geographic information. We use the term 'demonstration projects' to cover a spectrum of projects ranging from those where the main emphasis is on developing and showing new techniques − such as under the Alvey programme − to those which illustrate the broader operational aspects such as managerial and organisational problems. Projects such as the Dudley Metropolitan Borough Council LAMIS project have more or less fulfilled the functions of the latter sort of demonstration project.

7.19   We believe demonstration projects are a particularly effective means of illustrating applications and benefits of geographic information technology, provided they include the following key elements:

1   the application of the technology to a real operational activity;

2   the objective monitoring of costs and benefits; and

3   a commitment to disseminate the results of the project.

7.20   Setting up and funding such demonstration projects could most successfully be carried out through the collaboraton of users, system suppliers and data suppliers. The user involvement could in many cases be based on a group or club of users with similar interests so as to share costs and spread awareness. This approach might be particularly suitable for local authorities such as a group of 'urban' Borough or 'rural' District Councils faced with common needs. The more technically based demonstration projects might also involve academic researchers and be eligible for support funding from the UK Government and European support schemes for industry. Our proposed central body could coordinate the setting up of demonstration projects and help to monitor their costs and benefits and disseminate the results.

7.21   *We recommend that a limited number of demonstration projects based on real life applications should be set up and funded by consortia of users, and systems and data suppliers. The costs and benefits of the projects should be monitored and the results widely disseminated by the consortia. Our proposed central body should support and coordinate these activities.*

Demonstration of computer systems

7.22   The most successful way of demonstrating the effectiveness of particular Geographic Information Systems will be by way of the demonstration projects described above. However, it is unlikely that those projects would be able to meet all of the needs of potential users to acquaint themselves with particular equipment and software. We believe that further demonstration resources are required.

7.23   A few universities currently act as informal demonstration centres for this purpose, but their resources are already stretched. In the wider Information Technology area the Department of Trade & Industry (DTI) has been active in promoting demonstrations of equipment and software.

We recommend that ways of meeting the need for demonstrations of Geographic Information Systems equipment and software should be further explored by our proposed central body in collaboration with the DTI.

99

7.24 Many submissions of evidence to the Committee called for a source of impartial expert advice to help users select suitable equipment. Currently there is little independent advice on the capabilites of particular systems, including hardware and software. There are few consultants with the experience to provide advice. It is likely that this gap will quickly close as more consultants develop expertise in this field.

7.25 A potential source of advice for Government users is via its Information Technology support services for each Department. The Local Authorities' Management Services & Computer Committee (LAMSAC) and other bodies, such as the Chartered Institute of Public Finance & Accountancy (CIPFA), may be able to perform a similar role for local government users. Although such organisations are currently able to give advice on a range of Information Technology systems, this has not included Geographic Information Systems: their advice activities need to be strengthened in this respect.

7.26 *To encourage users to seek consultancy advice, we recommend that a register of consultants with experience of Geographic Information Systems should be set up. The proposed central body should consider how this could be done.*

8   Education and training

8.1   We said in Chapter 7 that the full exploitation of geographic information requires users at all levels to be aware of the relevance and benefits of developments both in general Information Technology and in geographic information technology. Education is essential to achieving this. Equally important is the existence of trained personnel at all levels, ranging from highly skilled developers of geographic information technology and systems to 'on the ground' operators of systems who need opportunities for training, both 'on the job' and on more formalised courses.

8.2   A shortage of trained personnel in this area was identified in the House of Lords Select Committee on Science and Technology report *Remote Sensing & Digital Mapping (1983),* and the report accordingly made a number of recommendations to remedy the situation. We note that, since then, provision of education and training opportunities for remote sensing has increased although a Royal Society working party has been set up recently to examine what further measures are required to improve provision. But there has been little increase in education and training provision for geographic information handling.

8.3   From evidence put to us, it is apparent that there continues to be a serious gap between education and training requirements and actual provision in the geographic information handling areas. In our view, this gap is a factor holding back the use of technology for handling spatial data and the shortage of trained personnel could become even more of a constraint in the future as demand increases. Given the long lead times to achieve improvements in educational provision, any measures taken to close this gap should be instigated as soon as possible. In this Chapter, we assess the improvements needed in general education and in specific training for geographic information handling.

General education   8.4   Attention has already been drawn to the overall lack of awareness in the United Kingdom of the potential application of Information Technology. The inclusion of Information Technology in general education is a vital step towards increasing awareness of it. It is also needed to provide a basis for more specialised training in later life. Educational requirements for the geographic information area mirror those for Information Technology.

8.5   Two recent reports − by the Government's Information Technology Advisory Panel (ITAP) *Learning to Live with Information Technology* and the Organisation, for Economic Cooperation & Development *New Information Technologies: A Challenge for Education* (Refs. 19 & 29) − were both critical of current educational provision for the electronic age. The Departments of Education & Science and Trade & Industry, and the Manpower Services Commission (MSC) are taking a number of measures to increase the use of Information Technology in schools. The ITAP report recommended further action to increase the application of Information Technology and demonstration of innovative projects throughout the education system. We consider that raising awareness of geographic information technology should be part of these measures.

8.6　There is currently little on computer handling of spatial data in schools curricula although the Domesday project could provide a valuable and timely vehicle for furthering interest in the appreciation of spatial data handling. (The Domesday project and some other teaching software packages are outlined in *Box 8A*). In further and higher education, geographic information technology is increasingly incorporated in traditional 'geographic' areas e.g. land surveying, but not in other areas such as computer science.

8.7　We support the current and proposed Government measures aimed at increasing awareness of Information Technology in schools and post school education.

More specifically, we recommend that, as part of these measures aimed at increasing awareness of Information Technology, familiarisation with geographic information technology should be encouraged throughout the education system.

8.8　We believe that it would be appropriate to include a general appreciation of geographic information, its technology and its end uses in relevant post school training courses. One area where the use and analysis of geographic information has particular application is as a business or resources management tool; it would therefore be worthwhile to include it in appropriate business and administrative studies courses.

We recommend that an appreciation of geographic information technology and applications should be included in relevant post school training courses.

Opportunities for training

Training within employment

8.9　If an organisation does not possess skilled staff of the right kind to manage and operate Geographic Information Systems, it must either recruit new staff or existing staff must be retrained. In the latter case 'on the job' or on-site training is essential for both project managers and operators to familiarise them with the new equipment. Some training is often given by the suppliers of the new system. In addition to 'on the job' training, staff may need to be sent on external courses which give them a broader training, for example in the concepts underlying techniques of analysis.

8.10　In the general field of Information Technology, the recent ACARD report on *Software: a vital key to UK competitiveness* (Ref. 11), particularly emphasised the crucial role of employers, both in adopting Information Technology and in providing appropriate training for their staff. It considered that employers were not providing enough training, either on-site or off-site. Another recent report by the Engineering Council (Ref. 21) identified the apparent reluctance of companies to provide further training, and their preference for poaching staff on the grounds of convenience and economy. Poaching does not and will not help to reduce the overall shortage of skills in general Information Technology and in geographic information technology. Convincing employers of their role in training staff will be an important function of the awareness measures outlined in Chapter 7.

8.11　We support the view that employers have a crucial role in providing more Information Technology training for their staff.

Box 8A: **Computer based teaching aids**

We outline below some of the few currently available computer based aids for increasing familiarity with Geographic Information Systems.

1 **The Domesday Video Disks:** The Domesday Project was initiated by the BBC to mark the 900th anniversary of the first Domesday Book. The project aims to present a contemporary snapshot of the United Kingdom in the 1980's on two interactive video disks and in a series of TV programmes. The material held on the disks includes 50,000 photographs and 250,000 pages of information. An innovative disk player has been developed by Philips to retrieve the data stored in the sound track and, under control of a micro-computer, combine data and images and display them.

The first or 'community' disk contains information collected by schools, community groups and individuals. The second or 'National' disk includes a selection of data from Government and quasi-Government sources. The system permits fast access to data, and allows interactive mapping and some limited data analysis by relatively unskilled users. The users of the system − which costs approximately £4,000 − are likely to be libraries, schools and local information centres such as tourist offices.

2 **The Advisory Unit for Computer Based Education (AUCBE):** The AUCBE develops, publishes and distributes the most widely used information handling software in UK schools and also provides support services, advice and in-service training courses. Examples of spatial data packages which the AUCBE has produced are:

1 1981 Census of Population data at district level: including population data and digitised boundaries of England and Wales. It can be used on a DEC mainframe and RML and Acorn BBC micro-computers.

2 1981 Census of Population enumeration district level tables: including population, social and economic data and digitised boundaries of each enumeration district, the main road network and main land use areas in Stevenage. It is available on floppy disk for RML and Acorn BBC micro-computers.

3 **The ARCDEMO Tutor:** This tutor was developed by Birkbeck College, University of London, to illustrate basic Geographic Information System concepts and functions. It is also available on-line to the academic community over the Joint Academic Network (JANET), the university computer network. It contains both text and multi-colour graphics. Operations covered include automatic data validation and correction, map projection change, selective retrieval of spatial data, map overlay, and network analysis.

Due to the success of ARCDEMO another system − ECDEMO − was designed to show how the Geographic Information Systems being used in the Coordinated Information on the European Environment (CORINE) project, referred to earlier (para. 2.46), integrated and mapped a variety of European data sets. The system has been used by others involved in the CORINE project, including coordinators in Europe, over international packet switched computer networks. The user text is offered in a number of European languages.

We recommend that employers should provide training opportunities in the handling of geographic information for managers and operators.

Post degree training

8.12   Several submissions of evidence − for example those from the HM Land Registry, Durham University and the Royal Societies − identified a shortage of personnel with the necessary capability to develop and manage Geographic Information Systems. Post graduate full-time courses of one or two years and PhD/research studentships are required to overcome this deficiency.

8.13   Provision of higher education diploma or masters courses related to geographic information has expanded within the last two to three years. There is now one course specifically for Geographic Information Systems. It is run by Edinburgh University and provides approximately 15 places annually. In addition, there are a number of diploma and masters courses, including those in remote sensing, which offer special options in geographic information. Demand from UK students for all these courses appears to be high; in addition, the courses attract many overseas students. Employer sponsorship of places on the courses is high and students who are not sponsored apparently have no difficulty obtaining employment on completion of their courses.

8.14   To accommodate the high level of demand, Edinburgh University has doubled the number of places offered on its recently established MSc course in Geographic Information Systems. However, there is a limit to the expansion of existing courses and the creation of new ones without provision of more staff, equipment and student grants. Within universities this depends substantially on what priority individual universities give to these courses. However, without overall expansion, the creation of new courses or the expansion of existing courses currently requires the reduction of places elsewhere in a university.

8.15   It is apparent from the evidence put to us that there is a serious shortage of up-to-date equipment in educational institutions. Specialist courses on handling geographic information require a relatively high level of equipment, including computers and software, to enable students to gain sufficient practical experience. This point was echoed in a number of submissions of evidence.

8.16   To secure and maintain appropriate hardware and software equipment, extra finance is often required to supplement that which is provided by the education institution. Some universities have secured this through building up strong consultancy activities. Edinburgh University considers that its many research activities undertaken on a consultancy basis were an essential prerequisite for its MSc course, both in providing funds and equipment and as examples of real world applications. However building up a consultancy base is usually a slow process: a critical constraint is the amount of staff time it requires.

8.17   It is often in the interest of systems suppliers to collaborate with universities and to provide educational discounts on their products, partly to identify promising people for future employment, partly to benefit from research activities, and also to familiarise future users with their products. This practice is common in North America: we hope that it becomes more widespread in the United Kingdom.

8.18   In addition to finance for equipment and staff, grants are required for students not sponsored by employers and for researchers. There are currently few grants available for these students. On the Edinburgh MSc course, for example, there are currently 4 Scottish Education Department grants available for Scottish students only and no grants for students from the rest of the United Kingdom. The only other national sources of student grants that we are aware of are a small number provided by the Natural Environment Research Council and the Economic & Social Research Council.

8.19 *To increase provision for post degree training in geographic information handling we recommend that:*

1 *The appropriate funding bodies for higher education institutes should give higher priority to providing new courses in handling geographic information and to expanding existing ones.*

2 *Higher education institutes should ensure that up-to-date equipment and software are available for courses using geographic information technology.*

3 *The Research Councils should consider increasing the number of postgraduate training awards in the handling of geographic information.*

Other training opportunites

8.20 We consider that technician training is important in providing a foundation for system operators who can develop their skills as necessary through further courses or on-site training. Although we received little evidence from practitioners in this area, it was put to us that there is still little technician training in further and higher education colleges. Although a number of courses use geographic information extensively, there is only one Business and Technician Education Council (BTEC) Higher National Diploma course, run by Luton College of Higher Education, on the practical aspects of geographic information handling. The funding, equipment and staffing problems referred to in relation to expansion of post degree training apply equally to this level of provision.

8.21 We noted the innovative provision by Brighton Polytechnic and the MSC of a skills training course in the handling of spatial data for unemployed graduates. However, there is generally a poor level of provision of short courses to enable personnel at all levels to update their knowledge or develop their skills. Securing funding for staffing of short courses within further and higher education institutes is particularly difficult.

8.22 The use of distance learning to improve skills has been developed particularly by the Open University and the National Computing Centre, and is being increasingly funded by the MSC. It is as yet undeveloped in the geographic information field. The IT'86 Committee's recent report (Ref. 12) emphasised the importance of developing interactive distance learning techniques for people with a relatively low level of skills. We consider that they are also of value to some professionally qualified people, for example those required to undertake continuing professional development by organisations such as the Royal Institution of Chartered Surveyors. A similar need is for computer based teaching packages for use in further and higher education institutes.

8.23 Despite a relatively large market, we are aware of only one interactive teaching package for geographic information handling. The MSC and the University Grants Committee Special Initiatives Scheme for Computer Based Learning may be appropriate sources of support for the development of interactive distance learning or teaching packages.

8.24 *We recommend that:*

1 *In addition to post degree courses the BTEC, MSC and other providers of training should make an assessment of requirements for other training courses in the handling of geographic information.*

2 *The Department of Education & Science and the MSC should encourage the development of computer-based interactive packages for teaching and home based learning in the handling of geographic information.*

Course content

8.25 To supply the necessary foundation for training courses specifically for geographic information handling, a strong case was put to us on the need for an interdisciplinary approach, including particularly geography, computer science and statistics. Yet training courses still tend to be based predominantly in geography departments.

We recommend that training institutes achieve more interdepartmental collaboration in the provision of specialist courses in geographic information handling.

Retraining and training teachers

8.26 The demand for education and training is increasing but there is currently only a small body of experts in the handling of geographic information. As in other Information Technology areas, it is difficult to retain existing teachers or to recruit new ones since remuneration is significantly lower than that offered by suppliers of hardware and software.

8.27 Given the rapidly changing technology and the increasing number of existing and potential applications, it is difficult for teachers to keep abreast of new technology developments and user requirements. The measures to increase awareness identified in Chapter 7 would help to overcome this problem and we have indicated there that we consider teachers to be an important target for such measures.

8.28 It was suggested by the Royal Society, the Royal Society of Edinburgh and the Royal Geographical Society, that a further way of overcoming this problem would be to encourage secondments between teachers, users and system suppliers. We are aware of the success of the Science & Engineering Research Council sponsored 'teaching company' scheme, which places teachers in companies to further the exploitation of research; we believe such secondments in the geographic information handling field would be fruitful.

Achieving improved provision

8.29 Achieving an improvement in education and training will require collaboration between employers, systems suppliers, educationalists and controllers of public education provision.

8.30 We are conscious that responsibility for public education provision lies with many different bodies and that there is no one body to take the lead in improving opportunities for education and training in geographic information handling. We believe that the improvement of education and training opportunities in remote sensing (referred to in para. 8.2), faced with a similar problem of divided responsibilities, was helped by the supporting activities of the National Remote Sensing Centre. We believe there is a need for similar action to help improve geographic information handling education and training provision. This could be carried out by the central body we propose in Chapter 10. It could help to assess requirements, review progress and encourage the development of opportunities in particular areas, such as in the provision of computer based teaching packages.

9 Research and development

9.1 In a field where the technology is relatively young and is developing rapidly and where people are still experimenting with how best to use it, it is not surprising that the evidence stressed the importance of research and development (R&D). The Royal Society put it concisely in stating 'R&D has a critical part to play in developing efficient and cost-effective methods of handling geographic information.'

9.2 We accept the need for R&D but believe it is important to be clear about the primary objective. This is, in our opinion, that the main R&D effort should be directed towards exploitable results. This will have the added advantage of providing UK suppliers with the opportunity to develop a healthy and competitive market in geographic information handling products and services. From our review, we have identified three general issues which bear on the primary objective.

9.3 First, a good deal of the research work which is relevant to our field is carried on as part of the wider field of research required for Information Technology generally. It is important to distinguish between this research and that which is specific to geographic information handling. But we note that some specific research may well have important spin-offs in other fields.

9.4 Second, a balance should be struck between research needed for current applications and that for the longer term. It is important to keep in mind long-term trends and opportunities − some basic research of an exploratory nature is needed to underpin future development. As the Information Handling Working Group of the National Remote Sensing Centre put it: 'applications cannot proceed unless basic research and development have been carried out, nor can research and development effort be directed rationally unless a detailed awareness of user needs exists'.

9.5 We therefore stress the importance of interaction between academic research activities and those of the commercial sector. We do not want to give the impression that current collaboration between the academic and industrial sectors is poor; indeed we have seen a number of examples of good interaction between the two.

9.6 We also stress the need for a healthy academic research base. This point was put to us forcefully by one of the leading North American practitioners, Tomlinson Associates, who said:

> '. . . past experience (shows) that most major research breakthroughs will occur in governments and universities rather than the private sector; commercial incentives appear to have led to successfully adapting and improving, but not to major new directions . . . '

9.7 Third, geographic information technology is an enabling technology which is often important but not central to the interests of users whose constituency is dispersed and difficult to bring to a focus. This problem arises particularly in the provision of public funding of research and we deal with it later in this Chapter.

**Where research is
required**

Obstacles to progress

9.8   The operation of Geographic Information Systems includes the initial capture of data in digital form, database management and understanding how the data can be reliably used. Each of these processes involves a variety of technologies, both hardware and software based; trends in their development have been outlined previously. Although rapid advancement in the technology has enormously broadened the potential uses, there are nevertheless a number of bottle-necks — which we outline below — impeding further evolution of the systems.

Data capture

9.9   We have already drawn attention to the massive task of converting data currently 'locked up' in analogue form. The take-up of Geographic Information Systems would be greatly influenced by the development of cheaper and faster data conversion techniques.

Database management

9.10   There is a need to manipulate very large volumes of data and to integrate different data sets whilst maintaining tolerable response times. This requires major developments in basic machine hardware, data structures, high density storage and efficient network facilities.

Data analysis and output

9.11   As spatial data becomes easier to access and manipulate, the scope for interpreting data will increase. Some of the problems involved in the statistical analysis of geographic information were illustrated in *Box 6A*. Existing statistical methodology which can be applied to spatial data is little developed: hence there is a need to develop this field.

More flexible systems

9.12   Users need to manipulate and analyse data in a more sophisticated way through spatial model building and asking 'what if' questions. This will require the development of Intelligent Knowledge-Based Systems (IKBS).

Research & Development tasks

9.13   To overcome these bottle-necks, the main R&D tasks which are needed and which are particular to geographic information handling, are set out below.

Data capture

9.14   Improvements are needed in automatic and semi-automatic cartographic data capture by raster scanning and line-following techniques. This will avoid bottle-necks in the generation of digital data.

Database management

9.15   The development of data structures suitable for holding all forms of spatial data compactly and allowing efficient retrieval and processing is required. One particular task in this area is the development of techniques for integrating raster data including remotely sensed data, and vector data.

Data analysis and output

9.16   Two main tasks in this area are:

1   The development of 'user friendly' software so that the ordinary user can operate systems without the need for expert assistance.

2   The development of statistical methodology for the linking and analysis of spatial data. Research is also needed into methods of defining data quality and to develop measures of the reliability of inferences that are drawn after combining multiple data sets.

9.17 These incorporate rules derived from human knowledge and experience but are at present in a rudimentary state of development. The potential benefits, even in the short term, are great and pervade most aspects of a Geographic Information System:

1 There is already promising work in progress on the automated recognition of features on maps. In map digitising, this may well mean that the lines can be coded with little or no human intervention. In remote sensing, methods of automatically classifying data require development by the incorporation of textural and contextual information. Both these areas require the application and development of work now being carried out in pattern recognition.

2 Even after the data are collected and coded, there are other tasks in which IKBS could be of great value. Perhaps the most fundamental is simplification − or generalisation − so that, for instance, the same Ordnance Survey (OS) data could be used to produce maps at widely differing scales; at present, this is impossible to achieve satisfactorily.

3 Other requirements include the selection of the best strategies to search a database for different kinds of queries and types of data.

4 In the longer term, such systems will need to have the capacity for learning from previous results and to provide assessments of the reliability of the results achieved. Beyond that stage, the ability to elicit and store expert but currently unrecorded spatial knowledge would have fundamental implications for our ability to understand and manage the environment.

5 We find it surprising that, for all the many years in which maps and other diagrams showing spatial data have been made, the principles of what is effective design are still a matter for argument. Nowhere is this more true than in relation to computer displays: the constraints imposed by the hardware mean that graphic design for visual display unit is very different from that for a general purpose paper map. This too would be a beneficial application of IKBS techniques, although significant preliminary work will be required on the human factors involved in understanding maps.

General developments in Information Technology

9.18 In three other fields we believe the research tasks can depend mainly on R&D likely to be carried out in the general field of Information Technology as follows:

1 The development of basic machine hardware with much higher processing speeds than are currently achievable from typical general purpose serial processing computers. These improvements will probably involve the use of Very Large-Scale Integrated Circuitry in transputers.

2 The development of higher density storage devices including enhanced methods of on-line archival storage for massive data volumes. The new devices will include high-density magnetic storage as well as media such as optical disks.

3 Efficient network facilities at a national and local level. User requirements will range from a quick-look capability to the dissemination of very large data sets and will dictate transfer rates according to task.

Funding

Role of Government

9.19 Government already has an established and valuable role, both in supporting basic research via academic institutions and the Research Councils and in encouraging commercial activity to exploit the results of that research through a number of support schemes. It also directly commissions research for its own operational activities.

Government support schemes

9.20 There are a number of UK and European Government schemes for developing innovative technology; we summarise the main ones in *Box 9A*. We note with approval that collaboration with academic institutions is often a condition of funding. One source of support for research involving remotely sensed data is the British National Space Centre.

9.21 Despite some views to the contrary, we see no good reasons for creating fresh support schemes to meet the R&D requirements of geographic information handling or for new mechanisms to help secure funds for this area from existing support schemes. Nevertheless, geographic information research has received little funding from existing schemes: this is unsatisfactory. We suspect that important factors have been the inability of this dispersed constituency to focus its needs (see para. 9.7), coupled with a lack of familiarity with this field by those administering the support schemes.

9.22 The report of the IT '86 Committee (Ref. 12) was published while we were preparing this report. Its analysis on the wider front of Information Technology parallels our own thinking. It recommended a plan of action based on a scheme of Information Technology Application Projects and a Research Effort. The Applications Projects would be collaborative projects between users and suppliers with the objective of stimulating the exploitation of Information Technology research. The Research Effort would be geared to future market opportunities and the development of Information Technology applications. It would consist of work generated from the Application Projects and additional longer term work on enabling technologies related to Information Technology systems.

9.23 A number of the research areas selected by the Committee will be relevant to our field. We also believe that the technology for Geographic Information Systems should be an essential project for the proposed Application Scheme. The breadth of the home market, the evidence of a large overseas market and current developments of applications all suggest that Geographic Information Systems meet the Committee's suggested criteria.

9.24 We support the Committee's recommendations, and the twin concepts of the Applications Scheme and the Research Effort and the argument that lies behind them.

We recommend that projects based on geographic information technology should be included in the Applications Scheme proposed by the IT'86 Committee.

Government's own R&D requirements

9.25 A number of Government Departments sponsor R&D of direct interest to their operational tasks – notably, where geographic information is concerned, the OS and the Ministry of Defence (MOD). In general, such work is highly targeted and the results are often unknown outside the sponsoring Department. This is particularly important in the case of the MOD because of the high expenditure on spatial data R&D for military command and control systems. We welcome its recent initiative to market its technology and products and hope that this approach will be taken up by other Departments.

Box 9A: Support schemes for research & development

A number of national and European schemes provide funding to support innovation and research and development activities.

Support for Innovation Schemes: Within its £160 million budget for support of general industrial R&D, the Department of Trade & Industry (DTI) offers financial assistance to encourage firms to undertake innovative projects which promise high returns. All sectors of technology are eligible but support is given selectively to those projects which are judged most likely to produce a significant advance. The main form of support is a grant of up to 25% of the eligible costs although collaborative projects involving two or more organisations may be eligible for grants of up to 50% of the relevant costs. Funding for demonstration projects is available where there is a need to establish wider market credibility for innovative and newly developed products.

The Alvey programme: This 5 year programme of R&D in advanced Information Technology was established in 1983 within a financial envelope of £350 million to which Government contributes more than half. Projects are jointly sponsored by the DTI, the Ministry of Defence, the Science & Engineering Research Council and industry. The objective is to improve the competitive position of UK industry to the world-wide Information Technology sector. The key feature is collaboration: projects normally involve at least two firms, together with academic institutions as appropriate. The industrial partners receive 50% of the costs of a project, academic partners being funded at the 100% level. The Alvey programme also supports large scale demonstrator projects aimed at applying the results of R&D to practical taks. Alvey funding is now fully committed and a follow-up programme is under consideration by Government on the basis of the report of the IT'86 Committee chaired by Sir Austin Bide.

LINK: This scheme, announced in December 1986, is aimed at speeding-up and increasing the commercial exploitation of scientific research undertaken by universities, Government Departments and industry. This initiative is expected to release expenditure by Government and industry of at least £400 million over the next five years. The Government's funding of £200 million will be mainly via existing arrangements for supporting R&D. All Government Departments having significant programmes are involved in LINK. Programmes or individual projects can cover the entire spectrum of technology and its applications. Each project will last 1-5 years, funded at least to 50% by industry over its lifetime, and will normally involve collaboration between users, industry and the academic science base. An interdepartmental LINK Steering Group has been formed to guide the implementation of the scheme.

EUREKA: This is an European framework aimed at stimulating collaborative R&D by industry to increase commercial competitiveness. To be accepted under EUREKA, the projects must involve industrial collaboration between at least two of the nineteen EUREKA countries. Current projects include high technology products for vehicles, homes and offices, advanced manufacturing and biotechnology. Financial support to UK companies is available at up to the 50% level through the DTI's Support for Innovation Scheme.

ESPRIT: The European Strategic Programme for Research and Development in Information Technology (ESPRIT) is the European Community's 10 year programme (1984-93) of collaborative research intended to provide European Information Technology industry with the technology to compete in world markets. The first five year phase is underway within a financial envelope of £1 billion, of which 50% is funded by the Community. ESPRIT concentrates on supporting advanced micro-electronics and software development. The primary application areas are in the fields of office automation and computer integrated manufacturing.

| Universities and polytechnics | 9.26 Support for research in geographic information by academic institutes is mainly a matter for the individual university or polytechnic, but both Government and the relevant committees, such as the University Grants Committee (UGC), can exert influence. We believe that the level of commitment at the university and polytechnic level is, with some notable exceptions, inadequate especially in computer science and engineering departments. It is noteworthy, however, that modest investment in remote sensing through the New Blood scheme has rapidly improved the amount and quality of teaching and research in that area. The UGC's Special Initiatives Fund might also be a potential source of support for spatial information research. |

We recommend that the Government should draw the attention of the relevant higher education funding authorities to the desirability of improving research funding in geographic information handling.

| The Research Councils | 9.27 The Research Councils collectively have a wide and fundamental role in funding basic and some applied research concerning the development of Geographic Information Systems. But there is fierce competition from other research areas and the subject of Geographic Information Systems has so far been unable to secure a significant portion of the funds, either in the form of studentship or grants. We understand that the total amount of funds for research in Geographic Information Systems received from the Natural Environment Research Council (NERC) − under its Remote Sensing Special Topic Research scheme − over the last three years has only amounted to about £65,000. We ask NERC to consider funding a future special topic on Geographic Information Systems. |

9.28 In addition, the Economic & Social Research Council (ESRC) funded Regional Research Laboratories (referred to in para. 5.40) have a modest total budget of £140,000 for the first eighteen months although we were encouraged to learn that, if the initial scheme is successful, the budget would be increased substantially. We welcome the ESRC funded Regional Research Laboratories as a small but promising initiative for increasing research and teaching resources in the geographic information handling area.

9.29 One obvious difficulty the subject has in competing for Research Council funds is that it does not, and cannot, fall within the single remit of any one Council. Similar but less extreme problems have been faced by remote sensing and digital mapping: these were met by setting up a Joint Research Committee in 1985 to coordinate the Research Council's interests in those areas. To date, however, it has concentrated very largely on remote sensing.

9.30 We believe that a joint research committee is a beneficial method of tackling what is likely to be a serious problem where, as in this case, the area of interest overlaps all five Research Councils. We believe that the present arrangements are not effective for geographic information handling.

We recommend that the Advisory Board for the Research Councils takes appropriate steps to establish a Committee which can be effective across the whole range of geographic information handling. The chairmanship should be held by the Research Council making the greatest commitment to the area.

| **Overall coordination of research & development** | 9.31 We described in para. 9.7 how funding proposals may fail because of dispersed interest. What is needed is a forum or focus for R&D activities and the allocation of funding, which achieves better coordination without stifling |

initiative and innovation. The right solution lies along two related lines: better communication of research activities and results, which we turn to next, and a machinery to interface with the various funding agencies we described above; this we deal with in Chapter 10.

Exploitation of Research & Development

9.32   We wish to emphasise the importance of an interlinking of basic research, applied research, and development and users' requirements to facilitate users' exploitation of the results of R&D. Funding mechanisms will help to achieve this — for example, through support for collaborative projects between commercial firms, academics and users. In addition, we believe there are a number of other valuable ways of improving exploitation.

9.33   We identified in Chapter 7 a number of measures to promote the use of Geographic Information Systems and we believe these should be used to disseminate the results of R&D. We emphasised the value of demonstrations and we see considerable scope for the use of demonstration projects in several of the R&D areas we have identified. These can be pursued in the context of the Government and Research Council support schemes outlined earlier in this Chapter. We also identified (para. 8.29) the use of secondments as a valuable means of promoting collaboration on R&D.

9.34   We consider that two further valuable means of improving exploitation are:

1   *Registers of research:* There are several broad registers of research, such as the British Expertise in Science & Technology (BEST) database on individual researchers and the Current Research in Britain (CRB) register of the British Library covering the science and technology area in general. However, current classification of research areas in such existing registers is not apposite to research projects into geographic information handling.

   We recommend that the Government examines the need either to alter one of the existing research registers or to set up a new research register to cover the field of geographic information handling research.

2   *Publicly funded research:* Notifying the research register of ongoing publicly funded research in geographic information handling and disseminating the results — for example, through published case studies or seminars — can be incorporated as a condition of funding.

   We recommend that notification of the research register and publication of research results should be standard conditions of all publicly funded research contracts in the geographic information handling field, to the extent that these conditions are compatible with national security or commercial confidentiality.

10 The role of Government and the machinery for coordination

10.1   We have referred several times in our report to the requirement for some form of central support or coordination to aid specific aspects of Geographic Information Systems development. In addition, the very diversity of users and uses suggests that there is a prima facie case for some form of national framework — for a 'pulling together' of the many strands — and that the argument is in reality about the extent and form that the framework should take.

10.2   A common theme of the evidence to us was that there needs to be greater coordination in the handling of geographic information and that the Government should take a lead in providing it. The Government's own view is contained in its reply to the House of Lords Select Committee report (Ref. 2). It said, 'The Government agrees .... that there should be a widespread discussion on how Geographic Information Systems should develop, and that there is a need for a forum or "club" of users to help coordinate and strengthen their interests'.

10.3   However, it is important to distinguish between those areas where greater coordination will aid the more effective use of geographic information, which must be the central objective, and areas of apparent 'untidiness' which may be only a reflection of a healthy and natural evolution of a rapidly developing technology. In the latter case formal coordination is often unnecessary although monitoring arrangements within an overall policy framework may be required.

10.4   In these circumstances, the task is to devise a framework which recognises the diversity and allows an interplay between users, suppliers, and Government. The Government inevitably has a major role to play. It is the major supplier and user of geographic information, and many actions and activities can be carried out only by Government. Moreover, a firm lead by the Government would, more than anything else, provide a powerful impetus towards the more effective use of geographic information.

10.5   In this Chapter, we assess what functions a central body might have and the form it might take, particularly in relation to Government. To do so we examine the role of Government and others in furthering the use of geographic information technology generally and, more specifically, in carrying out the measures we have previously identified as appropriate for a central body to implement.

The role of Government and its relations with others

10.6   The Government is already involved in almost all aspects of the handling of geographic information, and there are a number of reasons why it should play a stronger role.

Data supplier and user

10.7   The Ordnance Survey (OS) is the custodian of our national geodetic and topographic mapping archive. The various actions we have recommended range from direct actions by OS to collaborative or consultative arrangements with users. In this last regard, we have recommended (para. 4.33) that OS should set up a working group with the major users of basic scales digital maps. This would be additional to the existing consultative committees.

10.8 Government Departments — such as the Census Offices, Department of Transport and the Agricultural Departments — are also responsible for the majority of the non-cartographic national data sets. The Government has taken steps to promote the use of its information through the Tradeable Information Initiative. We believe it should put further effort into releasing unaggregated data — and, where necessary, minimally aggregated data — and into establishing data registers and other means of facilitating outside users' location of and access to its data.

10.9 As a major user of geographic information, the Government has its own interest in using Geographic Information Systems to aid decision making. We believe a central body could assist the Government's own internal Information Technology and management advisory services in this regard.

Standards
10.10 We have made a number of recommendations for the development and setting up of standards — for locational referencing, for the spatial units by which data are held and are released, and for the documentation of data — to assist in the linking and exchange of spatial data. A particular and important case in point is the need for 1991 Census of Population data to be available, subject to confidentiality, on a unit post code basis.

10.11 We consider that the Government, as the main supplier of national data sets, is in an unrivalled position to set national standards effectively by its own adoption of them. We therefore ask the Central Statistical Office, as the head of the Government's Statistical Service, and the Department of Trade & Industry (DTI) as coordinator of the Tradeable Information Initiative, to ensure a consistent approach and to encourage the development and adoption of our recommended standards. A central body could support the Government in this task and encourage others to adopt these standards.

Legislative actions
10.12 Apart from the amendments to copyright legislation already proposed by the Government, the main requirement is to bring England and Wales into line with Scotland in respect of disclosure of property ownership and transactions in property. These are essentially matters for Government decision.

Awareness and the provision of skilled personnel
10.13 In general the issues we have identified concerning awareness and the provision of skilled personnel are rather wider than the direct responsibility of Government although it is in a position to give a lead and exert influence. To carry matters forward requires, in our view, a national focus in the form of a central body: we do not think it would be efficient to divide responsibilities for activities which are mutually supportive.

Research & development
10.14 The Government influences the scale and direction of much research and development (R&D), either directly — for example, through OS and the Ministry of Defence (MOD) R&D activities — or indirectly, through existing funding programmes. Many of our recommendations are aimed at Government — particularly DTI — and the Research Councils; further action clearly lies with them. However, the opening up of DTI funded support schemes to applications in the geographic information field may not of itself bring forth the necessary proposals for R&D from the UK supply industry. To

some extent, such proposals will depend on the industry's awareness of and confidence in the long-term prospects for investment in this field. We hope that our recommendations will help to generate the necessary confidence but it is important that some central agency has responsibility for encouraging R&D proposals in the UK supply industry, for recommending corrective action if these are not forthcoming and for making strategic assessments of future needs.

Links within and outside Government

10.15   It is apparent that the Government has a number of direct interests and responsibilities in the field of geographic information handling. It is also apparent that responsibilities devolve on several different Departments; that some of the actions arising from our recommendations are of an executive nature, others require policy decisions and some require coordination within Government. It is clear to us that there is no obvious central Department which has a sufficiently wide interest in the field to lead other parts of Government in carrying these matters forward. Moreover, in many areas, effective progress will only be achieved by involving agencies outside Government. Government has an essential role to play but it must involve these other agencies; equally, they must involve Government.

Role of common interest groups

10.16   Geographic information handling covers a wide range of data and users, and there are already a number of groups or 'clubs of users' — such as the National Joint Utilities Group — which reflect those interests in particular areas. Such groups, established through an evolutionary 'bottom-up' process and based on common interests, are more likely to be truly representational and provide better communication channels to and from the 'grass roots' than organisations imposed from the top.

10.17   In addition, common interest groups have a potentially powerful role in the development of more effective geographic information techniques through:

1   identification and representation of sectional interests;

2   arranging consortia funding of developments appropriate to sectional user requirements;

3   acting as consultative bodies; and

4   promoting information exchange between users.

The form and functions of a central body

Support from the evidence

10.18   Whilst we support a 'bottom-up' approach through common interest groups, their different interests need to be brought to a focus. There was strong support in the evidence for the establishment of a central body to act as an umbrella for the common interest groups and to meet the needs for coordination. Although there were different ideas on the form of a central body, many organisations believed that the Government should take an active part in setting up the body. For example, the Natural Environment Research Council suggested that the Government should set up a National Programme Board — chaired by the Department of the Environment with members drawn from Government, the research community and industry, and should nominate and provide support for national coordinating agencies in key sectors. The OS placed even more emphasis on a Government role in this area, suggesting that the coordinating role be taken on by an existing Government body, possibly the

OS. The MOD also emphasised the need for a coordinating machinery and suggested that this might be located in the Cabinet Office.

10.19   International evidence also supports the principle of a body, supported by national government, to coordinate geographic information interests overall or in particular areas. For example, the Dutch Government Advisory Committee on Land Information (the RAVI), with representatives of local and central government, the utilities and the legal profession, has had great success in the implementation of land information systems. Another example is the French National Council for Geographic Information, reporting to the Minister of the Plan. Its aims include: promoting geographic information; developing relevant technologies; carrying out studies; making proposals for future national policy; expressing opinions on proposed legislation; and coordinating programmes. In the USA, the National Science Foundation has published proposals for a National Geographic Information and Analysis Center which will be concerned with advancing and promoting the theory, methods, and techniques of handling geographic information.

Our conclusion   10.20   Although the evidence supported the need for a coordinating body it left unresolved the body's relationship with Government in terms of constitution and funding. One option, for example, might be for it to be based within a Government Department and funded wholly by Government. This would place it in a strong position to influence the other Government Departments responsible for progressing particular aspects of geographic information handling. Another option might be for a central body to be in the form of an independent unit outside Government and dependent on subscriptions from users. The main advantages here would be that it could better reflect users' interests and would involve a smaller call on public funds.

10.21   Having considered such options we concluded that the central body should not be within a Government Department, but should be closely linked to Government through its membership and funding. This arrangement would most satisfactorily reflect our view that development in this area will be determined by user demand for Geographic Information Systems. It would also provide users and others outside Government with a relatively direct means of promoting the changes they wish Government to carry out. For convenience, we have labelled this body the Centre for Geographic Information (CGI).

10.22   The CGI should be given a clear remit to carry out particular functions, referred to elsewhere in this report, as follows:

1   to provide a focus and forum for common interest groups, or clubs;

2   to carry out and provide support for promotion of the use of geographic information technology, including promotional activities carried out within and outside the general education and training process;

3   to oversee progress and to submit proposals for developing national policy in the following areas:

• the availability of Government spatial data, the operation of data registers and arrangements for archiving of permanent data;

• the development of locational referencing, standard spatial units for holding and releasing data, the operation of the post code system and the development of data exchange standards (cartographic and non-cartographic);

- the assessment of education and training needs and provision of opportunities to meet them; and

- the identification of R&D needs and priorities, including advice to Government on bids for R&D funds.

10.23 We believe the CGI should have a Board composed partly of representatives of the various clubs and partly of independent members. It will be important that the Government Departments with major interests in the field are represented. Its Chairman should be a person of standing and energy appointed by Government. The CGI should have a Director and small number of staff to carry out its identified functions and service the Board. The post of Director, and that of the Chairman, will be of key importance in ensuring that the Centre gets off the ground.

10.24 We see the CGI's core activities outlined in para. 10.22 being funded primarily by subscriptions from users, including Government Departments, and suppliers. Any further activities − such as the provision of substantial training courses or commissioned tasks from Government and others − which are consistent with its impartial status would be undertaken on a self-funding basis. The CGI should be able to expand to the extent that these two forms of funding are successfully developed.

10.25 In the first years it would be a great help if the Centre were also to receive launch finance from the Government in addition to any annual subscriptions from Government Departments as users. But the test of whether the Centre meets real needs will be whether those who use it are prepared to support it financially. Accordingly, the aim should be for the launch funding to cease after, say, five years. An early task would be to produce a three year rolling strategic plan, within twelve months, for submission to Ministers.

10.26 We emphasise that the CGI should be a separately constituted body, and not a part of a Department. However, since the Centre would need to be accountable to Government in its early years when it received Government launch funding and would have a close on-going relationship with Government Departments, we suggest that the Minister responsible for the OS should be its point of contact in Government.

10.27 *We recommend that a central body, independent of Government, is set up to provide a focus and forum for common interest groups in the geographic information area, undertake promotional activities and review progress and submit proposals for developing national policy. Its members should be from all interested groups and it should maintain strong links with the Government.*

List of recommendations

Digital topographic mapping

1 Ordnance Survey (OS) and the major current users of OS basic scale mapping should collaborate in full scale tests in an operational environment of maps digitised to the specification developed by the National Joint Utilities Group (NJUG). *(para. 4.33)*

2 OS should convene a working group with the potential users of basic scales digital mapping to guide the development of an accelerated digital mapping programme. The working group's remit should include the specification(s), quality control, costs, funding arrangements and timetables, for creating both initial basic scales digital mapping and any subsequent upgrading required. *(4.33)*

3 OS should be prepared to switch its resources into a more rapid digital conversion programme if the NJUG specification, or something similar, can later be brought up to a further fuller specification without incurring unacceptable additional costs. *(4.33)*

4 Should a more rapid programme result in extra funding being required this should be found by the main users. *(4.33)*

5 The more rapid conversion programme should produce a unified digital database for Great Britain with the OS responsible for ensuring suitable standards of quality and consistency. *(4.33)*

6 The aim should be for conclusions on a more rapid programme and on the future full OS database structure to be reached no later than December 1987, with a decision being taken to proceed by April 1988. *(4.33)*

7 These conclusions should be agreed between OS and the main users. If OS and the main users do not agree, independent advice should be made available to Ministers before decisions are taken. *(4.33)*

8 OS should give details of the copyright and royalty arrangements for digitising of maps for a more rapid conversion programme when a decision is taken on the programme so that users and bureaux can have the necessary confidence to invest in staff and equipment. *(4.39)*

9 OS should publish guidelines explaining the principles on which it makes judgements on copyright and royalty arrangements to apply to new proposals for external digitising. *(4.39)*

10 External organisations should be able to negotiate terms with OS in advance for the marketing of digital data to third parties. OS should insist that such data is adequately documented as to its quality and limitations. *(4.39)*

11 OS should offer a much larger incentive to organisations prepared to digitise to OS standards where the resulting product can be of value to OS. *(4.39)*

12   OS should continue discussions with users to develop a charging basis which, whilst maintaining OS revenue, is considered by users to be fair. *(4.46)*

13   OS should consider whether it would be to its commercial advantage to encourage the development of digital mapping by reducing charges for digital mapping for wider research and educational uses, as it already does with its traditional products. *(4.46)*

14   OS, either itself or through agents, should sell the separate elements of its digital mapping at appropriate unit costs as soon as this is commercially viable. *(4.46)*

15   OS should continue to develop a better update service for digital mapping, and should consider introducing some reduction in charges for updated versions of a digital map to purchasers of an earlier version. *(4.50)*

16   Local authorities and utilities in Great Britain should agree procedures for the common recording of topographic information about new developments prior to the definitive survey by Ordnance Survey. This information should be available to Ordnance Survey as pre-survey intelligence. *(4.53)*

17   The OS longer-term objective should continue to be to achieve a scale-free digital database. *(4.63)*

18   OS should conduct new market surveys of the future demand for derived scales digital map data. *(4.63)*

19   OS should make an urgent assessment of the prospects for developing the software to produce 1:10,000 type data from the basic scales database. *(4.63)*

20   If the results of this assessment and the market surveys are positive, OS should support a research and development programme to develop the software to produce 1:10,000 type data from the basic scales database. *(4.63)*

21   Depending on the results of the market survey, OS should prepare and implement plans for conversion of the 1:50,000 map data to digital form. *(4.63)*

22   OS should not be expected to recover the full costs of providing digital data at derived scales in less than a 10 year period. *(4.63)*

23   To the extent that extra funds are required in the short term to implement these recommendations, this should not be at the expense of the basic scales conversion programme. *(4.63)*

24   The current restrictions on OS staff numbers and gross running costs should be lifted; a restriction on net costs and a recovery rate target are all that are required. *(4.64)*

Availability of data   25   Unaggregated spatial data held by Government Departments should be made available to other users provided that the costs of doing so are borne by the users and that there are no overriding security, privacy or commercial considerations. *(5.11)*

26　If unaggregated data cannot be released because of the need to protect confidentiality, minimally aggregated data should be provided on a standard basis. *(5.11)*

27　Treasury restrictions on gross running costs should be lifted where Departments can show a net return. *(5.15)*

28　Charges for data should be at marginal cost, and only at a higher rate if the market will bear it. For some users − such as educational establishments − reduced charges will be appropriate, as will be quid pro quo arrangements in some circumstances. *(5.15)*

29　Subject to confidentiality considerations, there should be increased use of franchising and other arrangements by which outside bodies can act as distributors of Government data. *(5.16)*

30　There should be open access to details of land ownership contained in Her Majesty's Land Registry's Register of Title and to details of land and property transactions held by the Valuation Office and the Valuation & Lands Office of Northern Ireland. This would bring England & Wales and Northern Ireland into line with Scotland. The necessary legislation to lift current restrictions should be introduced as soon as practicable. *(5.22)*

31　Legislation for the protection of digital data, as proposed recently by the Government, should be introduced as soon as possible. *(5.26)*

32　Government Departments should assess outside demand for their spatial data, particularly when computerisation of manual processes is to be introduced. *(5.31)*

33　Other public sector organisations should adopt the basic principles to increase the availability of spatial data which we have recommended that Government Departments should adopt. These are:

1　that unaggregated spatial data should be made available to other users subject to cost and confidentiality considerations;

2　that if unaggregated data cannot be released, minimally aggregated data should be provided; and

3　that measures should be taken to assess demand for data *(5.32)*.

34　Holders of spatial data collected at public expense should set up data registers individually or in groups as appropriate. *(5.37)*

35　The Central Statistical Office (CSO) should consider the setting up of one register for all Government Departments. *(5.37)*

36　The CSO and the British Library should consider the setting up of a central direction point to individual registers. *(5.37)*

Linking data sets　37　As far as practicable, all geographic information, including remotely sensed data, relating to the land areas of the United Kingdom should be referenced directly or indirectly to the National Grid or Irish Grid as appropriate. *(6.9)*

38　Data suppliers should both keep and release their data in as unaggregated a form a possible. *(6.11)*

39   The preferred bases for holding and/or releasing socio-economic data should be addresses and unit post codes. Wherever possible, the boundaries of administrative and electoral areas should not split whole unit post codes. *(6.22)*

40   All addresses and unit post codes should be Grid referenced. For addresses, the proposed provision of point references should meet most requirements. For unit post codes the current point references should be improved to meet users' requirements. To avoid different Grid references for the same address or post code, Government − probably the OS and/or the Census Offices − in conjunction with the Post Office, should be responsible for ensuring consistency in Grid referencing of addresses and unit post codes. *(6.26)*

41   The Office of Population Censuses & Surveys, in addition to the General Register Office (Scotland), should ensure that the results of the 1991 Census of Population and any future censuses are available, subject to confidentiality, on a unit post code basis. *(6.33)*

42   The Central Statistical Office and the Department of Trade & Industry should encourage and lead the development of data documentation standards within Government. Non-Government data suppliers should follow any Government lead in providing high standards of documentation for data sets. *(6.37)*

43   Maintenance of the adopted standards for the transfer of digital map data should not be a charge against OS resources, if OS undertakes this role. *(6.42)*

Awareness   44   Some central support should be established to provide a focus and encouragement for measures to promote the use and development of Geographic Information Systems. This should be in the form of the central body proposed in recommendation 64. *(7.14)*

45   User groups, data and system suppliers, and academics should actively disseminate information, via appropriate media, on the applications and benefits of Geographic Information Systems. *(7.16)*

46   More Geographic Information Systems familiarisation courses should be provided, perhaps on a joint funding basis by system and data suppliers and academic institutions. *(7.17)*

47   A limited number of demonstration projects based on real life applications should be set up and funded by consortia of users, systems and data suppliers. The costs and benefits of the projects should be monitored and the results widely disseminated by the consortia. The proposed central body should support and coordinate these activities. *(7.21)*

48   Ways of meeting the need for demonstrations of Geographic Information Systems equipment and software should be further explored by the proposed central body in collaboration with the Department of Trade & Industry. *(7.23)*

49   A register of consultants with experience of Geographic Information Systems should be set up. The proposed central body should consider how this could be done. *(7.26)*

Education & training

50 As part of the measures aimed at increasing awareness of Information Technology, familiarisation with geographic information technology should be encouraged throughout the education system. *(8.7)*

51 An appreciation of geographic information technology and applications should be included in relevant post school training courses. *(8.8)*

52 Employers should provide training opportunities in the handling of geographic information for managers and operators. *(8.11)*

53 The appropriate funding bodies for higher education institutes should give higher priority to providing new courses in handling geographic information and to expanding existing ones. *(8.19)*

54 Higher education institutes should ensure that up-to-date equipment and software are available for courses using geographic information technology. *(8.19)*

55 The Research Councils should consider increasing the number of postgraduate training awards in the handling of geographic information. *(8.19)*

56 In addition to post degree courses, the Business & Technician Education Council, the Manpower Services Commission (MSC) and other providers of training should make an assessment of requirements for other training courses in the handling of geographic information. *(8.24)*

57 The Department of Education & Science and the MSC should encourage the development of computer-based interactive packages for teaching and home based learning in the handling of geographic information. *(8.24)*

58 Training institutes should achieve more interdepartmental collaboration in the provision of specialist courses in geographic information handling. *(8.25)*

Research & development

59 Projects based on geographic information technology should be included in the Applications Scheme proposed by the IT'86 Committee. *(9.24)*

60 The Government should draw the attention of the relevant higher education funding authorities to the desirability of improving research funding in geographic information handling. *(9.26)*

61 The Advisory Board for the Research Councils should take appropriate steps to establish a Committee which can be effective across the whole range of geographic information handling. The chairmanship should be held by the Research Council making the greatest commitment to the area. *(9.30)*

62 The Government should examine the need, either to alter one of the existing research registers or to set up a new research register to cover the field of geographic information handling research. *(9.34)*

63 Notification to the research register and the publication of research results should be standard conditions of all publicly funded research contracts in the geographic information handling field, to the extent that these conditions are compatible with national security or commercial confidentiality. *(9.34)*

The role of Government and the machinery for coordination

64   A central body, independent of Government, should be set up to provide a focus and forum for common interest groups in the geographic information area, undertake promotional activities and review progress and submit proposals for developing national policy. Its members should be from all interested groups and it should maintain strong links with the Government. *(10.27)*

Costs of the recommendations

Many of the recommendations are concerned with the marketability of data held by OS, other Government Departments and public sector bodies and any short term increase in costs should be more than outweighed by the consequent increase in revenue. In the case of OS, if a more rapid programme of basic scales digital conversion requires extra funds these should be found by the main users. A derived scales digital coversion programme would require additional short term funding but this could be conditional on full cost recovery over ten years. The recommendations on Education and Training and Research and Development will not necessarily require additional funding if the money can be found by some modest redirection of funds within existing programmes. The main areas where some extra funding will be required are in launching the central body, the Centre for Geographic Information, and in its work on the development of awareness and the coordination of activity. The Centre will initially need about £250,000 a year from membership subscriptions, income generating activities and launch finance from Government. Such launch finance should taper to zero within 5 years.

References

1   *Remote Sensing & Digital Mapping:* Report of House of Lords Select Committee on Science & Technology: HMSO 1983

2   *Remote Sensing & Digital Mapping:* The Government's Reply to the First Report from the House of Lords Select Committee on Science & Technology: HMSO Cmnd.9320 1984

3   *Second Report of the Conveyancing Committee − Conveyancing Simplifications:* Prof. J T Farrand LLD: HMSO 1985

4   *Rural Wales Terrestrial Database (WALTER): Feasibility Study Final Report:* P M Mather and R H Haines-Young, University of Nottingham 1986

5   *Inequalities in Health:* Department of Health & Social Security 1980

6   *Fourth Report of the Steering Group on Health Services Information:* Chairman E Korner: Department of Health & Social Security 1984

7   *Neighbourhood Nursing − A Focus on Care:* Report of the Committee for Nursing Review − Headed by Julia Cumberledge: HMSO 1986

8   *NHS Management Enquiry Report: Review headed by Roy Griffiths:* Department of Health & Social Security 1983

9   *Roads and the Utilities − Review of the Public Utilities Street Works Act 1950:* Committee chaired by Prof. M R Horne OBE: HMSO 1985

10   *Geological Databank Pilot Study:* British Geological Survey, Natural Environment Research Council: 1986

11   *Software − A Vital Key to UK Competitiveness:* Cabinet Office Advisory Council for Applied Research & Development: HMSO 1986

12   *Information Technology − A plan for Concerted Action:* The Report of the IT '86 Committee: HMSO 1986

13   *Report of the Ordnance Survey Review Committee:* Chairman Sir David Serpell KCB, CMG, OBE: HMSO 1979

14   *Government Statistical Services: Review headed by Sir Derek Rayner:* HMSO Cmnd.8236 1981

15   *Government held Tradeable Information:* Guidelines for Government Departments in dealing with the Private Sector: Department of Trade & Industry 1986

16   *Property Law − Second Report on Land Registration:* Inspection of the Register: The Law Commission Report No.148: HMSO 1985

17   *Intellectual Property & Information:* HMSO Cmnd.9711 1986

18   *General Information System for Planning:* Report of Joint Local Authority, Scottish Development Department and Department of Environment Study Team: HMSO 1972

19 *Learning to Live with Information Technology:* Cabinet Office Information Technology Advisory Panel (ITAP): HMSO 1986

20 *New Information Technologies – A Challenge for Education:* OECD/SERI and HMSO 1986

21 *A Call to Action:* Engineering Council Report 1986

Abbreviations and acronyms

ABRC	Advisory Board for the Research Councils
ACARD	Advisory Council for Applied Research and Development
ACORN	A Classification of Residential Neighbourhoods based on Census statistics
AUCBE	Advisory Unit for Computer-Based Education
BEST	British Expertise in Science and Technology
BGS	British Geological Survey
BNSC	British National Space Centre
BSI	British Standards Institution
BSU	Basic Spatial Unit
BTEC	Business and Technician Education Council
BURISA	British Urban and Regional Information Systems Association
CAD/CAM	Computer Aided Design/Computer Aided Manufacturing
CGI	Centre for Geographic Information
CGIS	Canadian Geographic Information System
CHIPS	Computerised Highway Information and Planning System
CIPFA	Chartered Institute of Public Finance and Accountancy
CLUSTER	Central London Land Use and Employment Register
CORINE	Coordinated Information on the European Environment
CPD	Central Postcode Directory
CRB	Current Research in Britain
CSO	Central Statistical Office
DAFS	Department of Agriculture and Fisheries for Scotland
DBMS	Database Management System
DEmp	Department of Employment
DEn	Department of Energy
DES	Department of Education and Science
DHSS	Department of Health and Social Security
DMA	Defence Mapping Agency (USA)
D Mil Sy	Directorate of Military Survey
DOE	Department of the Environment
DTI	Department of Trade and Industry
DTp	Department of Transport
EC	European Community
ED	Enumeration District
ERS-1	ESA Remote-Sensing Satellite − 1
ESA	European Space Agency
ESPRIT	European Strategic Programme of Research and Development in Information Technology
ESRC	Economic and Social Research Council
GIMMS	Geographical Information Manipulation and Mapping System
GIS	Geographic Information System
GISP	General Information System for Planning
GKS	Graphical Kernal System
GRID	Global Resources Information Database
GRO(S)	General Register Office (Scotland)
HMLR	Her Majesty's Land Registry
HMSO	Her Majesty's Stationery Office
ICL	International Computers Limited

ICSU	International Council of Scientific Unions
IKBS	Intelligent Knowledge-Based System
ISO	International Organisation for Standardisation
IT	Information Technology
ITAP	Information Technology Advisory Panel
ITE	Institute of Terrestrial Ecology
JANET	Joint Academic Network
JIS	Joint Information Service (Tyne & Wear)
JUVOS	Joint Unemployment Vacancies Operating System
LA	Local Authority
LANDIS	Land Information System (MAFF)
LAMIS	Local Authority Management Information System
LAMSAC	Local Authorities' Management Services and Computer Committee
MAFF	Ministry of Agriculture Fisheries and Food
MARS	Merseyside Address Referencing System
MOD	Ministry of Defence
MOS	Marine Observations Satellite
MOSS	Modelling and Survey System
MSC	Manpower Services Commission
MSD	Master Survey Drawing
NAB	National Advisory Body
NATO	North Atlantic Treaty Organisation
NERC	Natural Environment Research Council
NJUG	National Joint Utilities Group
NOMIS	National On-line Manpower Information Service
NRSC	National Remote Sensing Centre
NSF	National Science Foundation (USA)
NTF	National Transfer Format
OECD	Organisation for Economic Cooperation and Development
OPCS	Office of Population Censuses and Surveys
OS	Ordnance Survey
OSI	Open Systems Interconnection
OSNI	Ordnance Survey of Northern Ireland
OSRC	Ordnance Survey Review Committee
PAF	Postcode Address File
PREMCODE	Premises Code
PUSWA	Public Utilities Street Works Act
R & D	Research and Development
RAVI	Dutch Government Advisory Committee on Land Information
RLUIS	Rural Land Use Information System
SASPAC	Small Area Statistics Software Package
SCDB	Sub-Compartment Data-Base
SDD	Scottish Development Department
SFI	Support for Innovation
SIF	Standard Interchange Format
SIM	Survey Information on Micro-Film
SPOT	Systeme Probatoire pour L'Observation de la Terre
SSSI	Sites of Special Scientific Interest
SUSI	Supply of Unpublished Survey Information
TRAMS	Transport Referencing and Mapping System
TRRL	Transport and Road Research Laboratory
UGC	University Grants Committee
UPC	Unit Post Code
UPRN	Unique Property Reference Number
USGS	United States Geological Survey

VO	Valuation Office
VOPREN	Valuation Office Property Reference Number
WALTER	Welsh Rural Areas Terrestrial Database
WDDES	World Digital Database for Environmental Sciences

Glossary of Terms

Accuracy The degree of conformity with a standard, whether absolute or relative. Accuracy relates to the quality of the result, and is distinguished from *precision* which relates to the quality of the operation by which the result is obtained. Higher accuracy implies that a measurement is nearer the truth.

Address A means of referencing a *property*, building or delivery point.

Algorithm A set of rules for solving a problem.

Analogue In the context of remote sensing and mapping, the term refers to information in graphic or pictorial form as opposed to digital form.

Archives Accessible store for historical records and data.

Attribute An attribute is a property of an *entity*, usually used to refer to a non-spatial qualification of a spatially referenced entity. For example, a descriptive code indicating what an entity represents, or how it should be portrayed.

Automated cartography The process of drawing maps with the aid of computer driven devices such as plotters and graphic displays. The term does not imply any information processing.

Automated digitising Conversion of a map to digital form using a method which involves little or no operator intervention during the *digitising* stage, for example, *scanning*.

Automated feature recognition The identification of map-based features using computer software incorporating pattern recognition techniques.

Cadastre The public register of the quantity, value and ownership of the land of a country.

Cartographic databank A store of data for drawing maps using computer aided cartographic techniques.

Cell The basic element of spatial information in the raster *(grid square)* description of spatial entities.

Centroid The point located within a *polygon* which coincides with the centre of gravity of a uniform sheet having the same shape.

Computer assisted cartography Production of maps using a computer to undertake all or part of the work.

Contour A line connecting points of equal elevation.

Databank A collection of data, in a common location, relating to a given set of subjects.

Database An organised, integrated collection of data stored so as to be capable of use by relevant applications with the data being accessed by different logical paths. Theoretically it is application-independent, but in reality it is rarely so.

Data capture The encoding of data. In the context of digital mapping this includes map *digitising*, direct recording by electronic survey instruments, and the encoding of text and *attributes* by whatever means.

Data compression	Methods of encoding data which reduce the overall data volume. (See *run length encoding*)
Database Management System (DBMS)	A collection of software for organising the information in a database. Typically a DBMS contains routines for data input, verification, storage, retrieval and combination.
Data model	An abstraction of the real world which incorporates only those properties thought to be relevant to the application or applications at hand. The data model would normally define specific groups of *entities*, and their *attributes* and the relationships between these entities. A data model is independent of a computer system and its associated *data structures*. A map is one example of an *analogue data* model.
Data register	A catalogue which provides information relating to *datasets*. The information may include − source, location, content, availability, revision date, etc.
Dataset	A named collection of logically related *features* arranged in a prescribed manner, for example, all water features. A dataset has more internal structure than a *layer* and is related to another dataset only by position.
Data structure	The logical arrangement of data as used by a system for data management; a representation of *data model* in computer form.
Derived scale map	A map produced from basic scale maps at larger scale.
Digital elevation model	A digital representation of a surface.
Digital ground model	A digital representation of relief (ground surface). Similar to a *digital elevation model*, but often enhanced by the addition of planimetric information.
Digital map data	The digital data required to represent a map.
Digital terrain model	Synonymous with *digital ground model*.
Digitising	The process of converting *analogue* maps and other sources to a computer readable form.
Encoding	The process of converting information to a computer readable form, for example, *digitising* maps.
Entity	Something about which data is stored in a *databank* or *database*, for example, a building or a tree. The data may consist of relationships, *attributes*, positional and shape information etc.
Epidemiology	The study of the spread of disease throughout populations in particular areas.
Feature	Another word for *entity*, commonly used in *cartographic/topographic databank/databases*.
Feature code	An alphanumeric code which is an *attribute* of a *feature* and describes and/or classifies that *feature*.
Format	The specified arrangement of data. For example, the layout of a printed document, the arrangement of the parts of a computer instruction or the arrangement of data in a record.
Geographic Information	Information which can be related to a location (defined in terms of point, area or volume) on the Earth, particularly information on natural phenomena, cultural and human resources. A special case of *spatial information*.

131

Geographic Information System	A system for capturing, storing, checking, integrating, manipulating, analysing and displaying data which are spatially referenced to the Earth. This is normally considered to involve a spatially referenced computer *database* and appropriate applications software.
Gigabyte	One thousand million bytes.
Grid	A planimetric frame of reference, for example the *National Grid*.
Grid squares	A regular array of rectangular cells referenced to a grid, used as a basis for holding spatially referenced information.
Hard copy	A print or plot of output data on paper or some other tangible medium.
Intelligent Knowledge-Based System	An interactive computer system, using programming techniques of artificial intelligence, designed for efficient problem solving and answering queries on the basis of knowledge acquired from experts.
Interactive digitising	A method of *digitising* in which dialogue (interaction) takes places between the operator and the computer.
Land information system	A system for capturing, storing, checking, integrating, manipulating, analysing and displaying data about land and its use, ownership, development, etc. Considered by some authorities to be a subset of a *Geographic Information System*.
Land parcel	An area of land, usually with some implication for land ownership or land use.
Locational reference	The means by which information can be related to a specific spatial position or location.
Manual digitising	A method of *digitising* which is accomplished by an operator moving a *cursor* over a map on a *digitising table*.
Map generalisation	The process of reducing detail on a map as a consequence of reducing the map scale.
The National Grid	The metric *grid* on a Transverse Mercator Projection used by the Ordnance Survey on all post-war mapping to provide an unambiguous spatial reference in Great Britain for any place or *entity* whatever the map scale.
Node	The start or end of a link or line. A node can be shared by several lines.
Optical disc	A compact data storage device consisting of a disc whose coating can be altered to encode information. Data can be read from the disc by laser light which is reflected from the surface according to the new property of the coating.
Parallel processor	An array of computing elements forming a data processor which can execute instructions on all elements of the array simultaneously.
Parcel	See *land parcel*.
Pixel	One of a regular array of cells in a collection of *raster data*, a picture element.
Point digitising	A method of *digitising* in which a cursor is placed over any position which is to be recorded and a button pressed to send that position to the computer. Contrasted with *stream digitising*.
Polygon	The piece of land, surface, or region bounded by a closed line or a closed series of lines.
Postcode	See *unit postcode*.
Precision	The exactness with which a value is expressed, whether the value be right or wrong.

Premcode	A set of alphanumeric characters which, when added to the *Unit Postcode*, serves to identify a specific premises or address.
Property	Land, building or estate which is owned and to which legal title exists.
Quadtree	A data structure for thematic information in a raster *database* that seeks to minimise data storage.
Raster data	Data expressed as an array of *pixels*, with spatial position implicit in the ordering of the pixels.
Raster map	A map encoded in the form of a regular array of cells.
Raster to vector	The process of converting an image made up of *cells* into one described by lines and *polygons*.
Remote sensing	The technique of obtaining data about the environment and the surface of the Earth from a distance, for example, from aircraft or satellites.
Representative point	A point with a *polygon* that can be used to carry the *attributes* of the whole polygon, for example, ownership, land use type. Also called area seed, peg point, point label.
Resolution	A measure of the ability to detect quantities. High resolution implies a high degree of discrimination but has no implication as to *accuracy*.
Revision	See *update*.
Run length encoding	A method of compacting *raster data* by recording counts of multiple, consecutive occurrences of a value, rather than repeatedly recording all the values.
Scanning	A method of *digitising* whereby an image or map is automatically detected and most of its *attributes* are converted into digital data in raster form. This is the method most commonly used for capturing remotely sensed data.
Semi-automatic digitising	A method of *digitising* from a map in which the majority of the line following is controlled by a machine, but which requires an operator to be on hand constantly to assist the machine.
Spatial information	Information which includes a reference to a two or three dimensional position in space as one of its *attributes*.
Thematic map	Map depicting one or more specific themes e.g. land classification, population density, rainfall etc.
Topographic database	A *database* in which data relating to the physical features and boundaries on the Earth's surface is held.
Topology	The study of the properties of a geometric figure which are not dependent on position; for example connectivity and relationships between *lines, nodes and polygons*. The way in which geographical elements are linked together.
Transfer format	The *format* used to transfer data between computer systems. In general usage this can refer not only to the organisation of data, but also to the associated information, such as *attribute codes* which are required in order to successfully complete the transfer.
Transfer medium	The physical medium on which digital data is transferred from one computer system to another; for example, magnetic tape.
Transputer	A term, derived from **trans**istor and com**puter**, which describes a programmable micro-computer and associated data-handling components all mounted on a single semi-conductor chip.
Unit Postcode	An alphanumeric code applying to a group of *addresses* (typically 15 in number) of neighbouring properties. Used in the sorting and delivery of mail.

Update The process of adding to and revising existing information to take account of change.

Vector data Positional data in the form of co-ordinates of the ends of *line segments*, points, text position etc.

Very Large-Scale Integration The process of designing and making devices consisting of very large numbers (typically 100,000) of data handling elements on a single semi-conductor chip.

Appendices

Appendix 1
Terms of reference, membership of the committee and the call for evidence

To advise the Secretary of State for the Environment within 2 years on the future handling of geographic information in the United Kingdom taking account of modern developments in Information Technology and of market need.

MEMBERSHIP
Lord Chorley (Chairman)
Mr W P Smith (Deputy Chairman)
Dr Jean Balfour
Mr I Gilfoyle
Professor D Rhind
Dr M Richardson
Mr F Russell
Professor R Schiller
Mr G Singh
Mr D Sinker
Dr J Townshend

Mr D Wroe, Director of Statistics, Department of the Environment was Assessor to the Committee.

SECRETARIAT
Mr J Metcalf
Ms C Myers
Mr J Troughton

Call for evidence
Interested organisations and individuals were invited to provide evidence for consideration by the Committee. The call for evidence listed a number of specific topics and these are set out below:

SECTION A. THE PRESENT POSITION
The Committee is aware of the wide range of activity included within its remit and sees its first task as that of taking stock of current information holdings whether in cartographic or statistical form, and whether in manual or computer based systems.

Evidence is therefore particularly sought on:

1 the main types of information you collect or use which are either spatially referenced or capable of being so referenced, (ie. they relate, or can be related, to specific locations) and the reasons (including statutory purposes) why the information is required;

2 the methods of collection, handling and updating;

3 the size of the information holdings and the extent of their geographical coverage;

4 the spatial units and referencing systems used (eg. national grid reference, post codes, postal address) and why these were chosen;

5 the extent to which the information is available to other users (a) within your organisation, (b) outside it, and under what conditions;

6 the extent to which you use allied information from other sources and relate this to your own spatially referenced holdings of information;

7 your policy for the retention of non-current information;

8 the lines along which improvements or development are currently being undertaken.

It would be helpful if you could provide separate evidence for each of the main types of information which you collect or use.

SECTION B. ISSUES FOR THE FUTURE
The Committee seeks the views of users and providers of information on major issues which may be relevant to the enhancement of information handling and systems, and on the nature of expected benefits to them which may justify the costs involved. Such issues may include but are not limited to:

1 What do you see as your needs for spatially related information over the next 5-10 years? Can relative priorities be indicated?

2 What are the main technological changes that will assist development in the future? In what areas do you hope that technology changes might materially increase efficiency or reduce costs?

3 What benefit do you foresee from linking or sharing data sets and what are the main constraints? Would on-line access to particular data sets be useful? Would you welcome access to information on, for example, underground utilities, property or planning data?

4 What do you consider to be the main areas of undesirable duplication in the collection and management of spatially related information?

5 What problems are there is using spatially referenced information currently supplied to you by other organisations and how might these problems be overcome?

6 Your views, as users or providers of information, on (a) data security, (b) confidentiality, (c) copyright, and (d) levels and methods of charging?

7 For your organisation, what are the arguments for either centrally held or locally distributed information systems, bearing in mind:

   (a) possible developments in nationwide networks for transmission of data;

   (b) needs for national and international standards for location-referencing and data exchange, and

   (c) the diversity of requirements among users?

8 What organisational arrangements are necessary for coordinating developments in this field? How might these developments be promoted?

9 Has central government a role in the development of geographic information systems? If so, what?

10 What do you see as the trade-offs for your activities between accuracy, completeness of coverage and currency of data on the one hand, and cost on the other hand? Please give examples.

11 What Research and Development (R&D) work are you planning to carry out in the future?

12 What R & D work should be undertaken in future by educational institutions, Government Departments, Research Councils, instrument and software firms and other public and private sector organisations? How should such R & D work be funded?

13 To what extent are educational and training needs being met and what action, if any, is needed for the future?

14 What impact do you expect a younger generation, more familiar with computers, to have on the way in which organisations deal with spatially referenced data?

15 What attention should be paid to developments in other countries and are there any lessons to be learnt about the way in which these matters are tackled elsewhere?

Background to the enquiry

The setting up of the Committee followed a recommendation of the House of Lords Select Committee on Science and Technology in its first report on *Remote Sensing and Digital Mapping* (Dec 1983). Extracts from their report are given below.

'5.11.3 What must be generally accepted is the need for improvement in methods for making data readily and conveniently accessible to users, integrating geographical databases and avoiding duplication of effort between data collectors. The communications network proposed by the Committee in section 5.3 will begin to provide a framework within which spatial databases can be created but widespread discussion between the interested parties will have to take place before decisions on geographic information systems can be made

5.11.4 For this purpose the Committee recommend that the Secretary of State for the Environment should appoint a Committee of Enquiry to report on the handling of geographic information in the UK, including the case for a national geographic information system, identifying administrative, financial and research responsibilities in this field. It should be an immediate enquiry, leading to a report within 2 years, and the Committee should have an expert chairman. The enquiry should have in mind the desirability of integrating remote sensing and other sources of spatial data in the same network.

Recommendation 28. The Secretary of State for the Environment should appoint a Committee of Enquiry to report on the handling of geographic information in the UK to report within 2 years.'

The Government in its response to the Select Committee's report accepted the recommendation for a Committee of Enquiry (Cmnd 9320 July 1984) in the following terms:

'37 As regards Recommendation 28 the Government agree with the Select Committee's view that there should be widespread discussion on how geographic information systems should develop, and that there is a need for a forum or 'club' of users to help co-ordinate and strengthen their interests. Experiments are already in hand, particularly with digital mapping data, to explore how the information might best be put to use, and this programme can usefully be extended. The experience gained in this way will need to be drawn together and used to ensure that further development of the techniques, and the ways in which the data is made available, respond to what users will require. Only with the public and private user's standpoint clearly represented will it be possible to achieve a practicable programme which ensures that the potential of the data is realised, and that the costs of the programme are in step with the expected benefits.

38 The Government therefore accept that it is timely to set up a Committee of Enquiry, as recommended by the Select Committee, to report on the handling of geographic information, including the preparation of a catalogue of all digital spatial data (see para 45 below). That Committee will also be expected to advise on the most appropriate framework for the representation of user interests in the longer term. A further annoucement will be made on the terms of reference, composition, etc of the Committee, following discussions with interested parties.'

Appendix 2
Organisations who submitted written evidence; organisations who submitted oral evidence; visits; studies

Organisations who submitted written evidence

A. CENTRAL GOVERNMENT DEPARTMENTS
Central Statistical Office
Department of Education and Science
Department of Employment
Department of Energy
Department of the Environment
Property Services Agency
Department of the Environment Northern Ireland
Ordnance Survey (NI)
Department of Health & Social Security
Department of Trade & Industry
Business Statistics Office
Warren Spring Laboratory
Department of Transport
Transport and Road Research Laboratory
Forestry Commission
General Register Office (Scotland)
Home Office
HM Land Registry
Inland Revenue (Valuation Office)
Ministry of Agriculture, Fisheries and Food
Ministry of Defence
   Hydrographic Department
   Meteorological Office
   Military Survey
National Remote Sensing Centre
Office of Population Censuses & Surveys
Ordnance Survey
Oversears Development Administration
Registers of Scotland
Scottish Office
Welsh Office

B. OTHER PUBLIC BODIES
Ancient Monuments Board for Wales
Boundary Commission for England and Wales
Boundary Commission for Scotland
British Committee for Map Information and Catalogue
   Systems
British Geological Survey
British Library
British Railways Board
British Steel Corporation
British Tourist Authority
Civil Aviation Authority
Countryside Commission
Countryside Commission for Scotland
Economic and Social Research Council (ESRC)
ESRC Data Archive
English Tourist Board
Estates Surveyors Group
English Heritage
Local Government Boundary Commission for England
Local Government Boundary Commission (Scotland)
Local Government Boundary Commission (Wales)
London Regional Transport
Macaulay Institute for Soil Research
Manchester Area Council for Clean Air
Manpower Services Commission
Medical Research Council
Metropolitan Police
Museums and Galleries Commission
National Coal Board
National Economic Development Office
National Remote Sensing Centre
   (Information Handling Working Group)
Natural Environment Research Council (NERC)
Nature Conservancy Council
Oxford Regional Health Authority
Peak District National Park
Royal Commission on Ancient and Historical Monuments
   (England)
Royal Commission on Ancient and Historical Monuments
   (Scotland)
Royal Commission on Ancient and Historical Monuments
   (Wales)
Royal Commission on Environmental Pollution
Rutherford Appleton Laboratory
Scottish Health Service
Soil Survey of England and Wales
Sports Council
West Midlands Regional Health Authority

C. UTILITIES
BBC
British Gas Corporation
British Telecom
Central Electricity Generating Board
East Midlands Electricity Board
South of Scotland Electricity Board
South Western Electricity Board
Yorkshire Electricity Board
Electricity Council
National Joint Utilities Group (NJUG)
North West Water Authority
South West Water Authority
Thames Water Authority
Wessex Water Authority
Water Authorities Association
Post Office

D. ACADEMIC INSTITUTES
University of Aberdeen
Advisory Unit for Computer Based Education
Aston University

Birkbeck College, University of London
Brighton Polytechnic, Countryside Research Unit
University of Bristol, School for Advanced Urban Studies
City of London Polytechnic
University of Dundee
University of Durham
University of East Anglia
University of Edinburgh
University of Essex
University of Glasgow
Imperial College of Science and Technology, University of London
University of Keele
Kings College, University of London
University of Lancaster
University of Leeds
University of Liverpool
Liverpool Polytechnic
London School of Economics, University of London
Luton College of Higher Education
University of Manchester
University of Manchester Institute of Science and Technology
University of Newcastle – Centre for Urban and Regional Development Studies
NE London Polytechnic
University of Nottingham
Open University
Plymouth Polytechnic
Portsmouth Polytechnic
Queen Mary College, University of London
Royal Holloway and New Bedford College, University of London
University of Reading
University of Salford
University of Sheffield
University of Southampton
University of Stirling
University College, Swansea
University College, University of London
University of Wales, Institute of Science and Technology

E. PRIVATE SECTOR ORGANISATIONS

Asda
W A Atkins Group
Automobile Association
Axis Software Systems Ltd
BKS Surveys
Bournemouth District Water Company
Britoil
Dr R Bullard
CACI
Cambridge Water Company
Collins Publishers
Colne Valley Water Company
Construction Industry Research and Information Association
Council for the Protection of Rural Wales
GEC Research Ltd
General Technology Systems Ltd
Geoplan UK Ltd
Gerald Eve & Co
Halifax Buidling Society
Healey & Baker
Hillier Parker May & Rowden
Intergraph
ICL

Jones Lang Wootten
Laserscan Labs Ltd
Leo Orenstein Associates
Logica Space & Defence Systems Ltd
G F Marshall Computer Services
Mid-Southern Water Company
National Westminster Bank Plc
Pinpoint Analysis Ltd
P M A Consultants Ltd
Plessey Defence Systems
Property Intelligence Ltd
Regional Newspaper Advertising Bureau
Racal Surveys Ltd
Royal Automobile Club
SysScan Ltd
Terrafix Ltd
Tesco Stores Ltd
Water Research Centre
Wilson Smith Residential
Wootton Jeffreys Consultants Ltd

F. LOCAL GOVERNMENT ORGANISATIONS

Association of County Archaeological Officers
Association of Chief Police Officers
Association of County Councils
Association of District Councils
Association of LA Valuers & Estate Surveyors
Association of London Borough Planning Officers
Association of Metropolitan Authorities
Association of Municipal Engineers
Convention of Scottish Local Authorities
County Planning Officers Society
County Surveyors Society
Local Authorities Management Services and Computer Committee (LAMSAC)
Local Authorities OS Committee
London Geographic Database Steering Group

G. SHIRE COUNTY COUNCILS

Avon
Buckinghamshire
Cambridgeshire
Cheshire
Cleveland
Clwyd (W)
Cornwall
Cumbria
Derbyshire
Devon
Dorset
East Sussex
Gloucestershire
Hampshire
Hereford and Worcester
Hertfordshire
Humberside
Kent
Lancashire
Mid Glamorgan (W)
Norfolk
Northumberland
Powys (W)
Staffordshire
Suffolk

Surrey
Warwickshire
West Glamorgan (W)
West Sussex
Wiltshire

H. METROPOLITAN AUTHORITIES
Metropolitan Counties
Merseyside
South Yorkshire
Tyne and Wear
West Yorkshire

London Authorities
GLC

Inner London
Camden
Hackney
Lambeth
Southwark
Tower Hamlets

Outer London
Barnet
Bromley
Enfield
Haringey
Hillingdon
Hounslow
Newham

Metropolitan Districts
Barnsley
Dudley
Kirklees
Manchester
Salford
South Tyneside
Wigan

I. DISTRICT COUNCILS
Aylesbury Vale
Bournemouth (B)
Bracknell
Cardiff (W) City
Carrick
Cheltenham (B)
Corby
Dacorum
Derby City
Durham City
East Devon
Epping Forest
Exeter City
Forest Heath
Gedling (B)
Great Grimsby (B)
Guildford (B)
Harborough
Hart
Isle of Angelsey
Kerrier
Kettering (B)
Kingston upon Hull City
Leominster
Luton (B)
Middlesbrough (B)

Mole Valley
Neath (WB)
Newbury
North Bedfordshire (B)
North Shropshire
Poole (B)
Purbeck
Richmondshire
St Albans City
St Edmundsbury (B)
Sedgemoor
Shrewsbury and Atcham (B)
South Cambridgeshire
South Oxfordshire
Stafford (B)
Suffolk Coastal
Thamesdown (B)
Thurrock (B)
West Devon (B)
Wimborne
Winchester City
Woodspring

J. SCOTTISH LOCAL AUTHORITIES

Regional and Island Councils
Regional Councils
Borders
Central
Dumfries and Galloway
Fife
Grampian
Highland
Lothian
Strathclyde
Tayside

Scottish District Councils
Aberdeen City of
Angus
Banff and Buchan
Berwickshire
Caithness
Clackmannan
Clydebank
Cumnock and Doon Valley
Cunninghame
Dumbarton
Dunfermline
East Lothian
Eastwood
Edinburgh City
Falkirk
Glasgow City
Gordon
Inverclyde
Kilmarnock and Loudoun
Kirkcaldy
Kyle and Carrick
Lochaber
Midlothian
Monklands
Motherwell
Renfrew
Perth & Kinross
Ross and Cromarty
Stirling

Strathkelvin
Tweeddale
West Lothian
Wigtown

K. PROFESSIONAL AND COMMERCIAL ASSOCIATIONS AND SOCIETIES
British Academy
British Ports Association
Cable Television Association of GB
Deep Earth Club
Faculty of Architects & Surveyors
Geographical Association
Guild of Surveyors
House Builders Federation
Institute of British Geographers
Institution of Civil Engineers
Institution of Geologists
The Law Society
Photogrammetric Society
Ramblers Association
Remote Sensing Society
Royal Geographical Society
Royal Institute of British Architects
Royal Institution of Chartered Surveyors
Royal Society
Royal Society — OS Scientific Committee
Royal Society of Edinburgh
Royal Statistical Society
Royal Town Planning Institute
Society of Surveying Technicians
Standing Committee of Professional Map Users
Water Companies Association

L. INTERNATIONAL
British Embassy, Washington
Bureau of the Census, Washington DC
Commonwealth Scientific and Industrial Research
    Organisation, Canberra Australia
Canada Land Data System, Ottawa
European Association of Remote Sensing Laboratories
International Geographical Union/International Cartographic
    Association/Joint Working Group on a World Digital
    Topographic Database
Urban and Regional Informations Systems Association,
    Bethesda, USA
US Department of the Interior
    Bureau of Land Management
    Geological Survey

Belgium — Institut Géographique National
Cyprus — Ministry of the Interior, Department of Lands and
    Surveys
Denmark — Geodaetisk Institut, Topografisk, Afdeling
France — Conseil National de l'Information Géographique
Federal Republic of Germany
    — Landesvermessungsamt Nordrhein -westfalen
    — Institut für Angewandte Geodäsie
Ireland — Ordnance Survey
Italy — Instituto Geografico Militare
Luxembourg — Administration de Cadastre et de la
    Topographie
Norway — Ministry of Environment
Portugal — Instituto Geográfico e Cadastral
Sweden — Lantmäteriverket — National Land Survey

Organisations who gave oral evidence
Central Statistical Office
Countryside Commission (England and Wales)
Countryside Commission (Scotland)
Department of Agriculture and Fisheries (Scotland)
Department of Employment
Department of the Environment
Department of Trade and Industry
Directorate of Military Survey, Ministry of Defence
Economic and Social Research Council
HM Land Registry
ICL
Local Authority Associations
Ministry of Agriculture, Fisheries and Food
National Joint Utilities Group
National Remote Sensing Centre
Nature Conservancy Council
Natural Environment Research Council
Ordnance Survey
Post Office
Royal Geographical Society
Royal Institution of Chartered Surveyors
Royal Society
Royal Society of Edinburgh
Royal Town Planning Institute
University of Bristol, School of Advanced Urban Studies
University of Newcastle, Centre for Urban and Regional
    Development Studies

Visits
Birkbeck College, University of London
BKS Surveys Ltd
British Gas, South East Region
Department of Agriculture and Fisheries for Scotland
Directorate of Military Survey, Ministry of Defence
Dudley Metropolitan Borough Council
ESRC Data Archive, University of Essex
Forestry Commission
General Register Office (Scotland)
Glasgow District Council
Grampian Regional Council
HM Land Registry
Hunting Surveys
ICL
Intergraph Ltd
NERC, Institute of Terrestrial Ecology, Bangor
Laserscan Ltd
Macaulay Institute of Soil Research
Merseyside Metropolitan Council
National Remote Sensing Centre
Natural Environment Research Council, Swindon
NJUG Digital Records Trial (Dudley)
Ordnance Survey
Ordnance Survey of Northern Ireland
Pinpoint Analysis Ltd
Registers of Scotland
South Oxfordshire District Council
Strathclyde Regional Council
SysScan Ltd
University of Edinburgh

Studies commissioned for the committee

ECOTEC	Market Demand Survey for Geographic Information in the Private Sector
Logica	Information Technology and Geographic Information Systems
Dr S Openshaw	Spatial Units and Locational Referencing
Tomlinson Associates	Review of North American Experience of Current and Potential Uses of Geographic Information Systems

The Committee also convened an International Seminar at the Royal Society on 22 September 1986. The Seminar was attended by

Professor J McLaughlin	Department of Surveying Engineering, University of New Brunswick, Canada.
Mr H Aalders	Cadastral Service of the Netherlands, The Netherlands.
Mr S Anderson	Director of the Central Board for Real Estate Data, Sweden.
Dr P Dale	Department of Land Surveying, North East London Polytechnic, London.
Professor G Eichhorn	Geodetic Institute, Darmstadt, Federal Republic of Germany.
Professor H Jerie	International Institute for Aerial Survey and Earth Science, Enschede, The Netherlands.
Mr Lowell Starr	Chief, National Mapping Division, US Department of the Interior, Geological Survey, USA.
Mr J Denègre	Secretary General, Conseil National de l'Information Geographique, France.
Professor I Williamson	Department of Surveying, University of Melbourne, Australia.

Appendix 3
Major data sets of spatially referenced information held by central government and local authorities

Sources	Datasets	Sources	Datasets
CENTRAL STATISTICAL OFFICE	Annual Abstract of Statistics Social Trends, Regional Trends	**DEPARTMENT OF THE ENVIRONMENT**	Land Register Database Urban Development Grants Urban Programme, Payments/ Expenditure
DEPARTMENT OF EDUCATION AND SCIENCE	Main Mechanised Record of Teachers Survey of Induction and In-service Training of Teachers Destination Survey Secondary School Staffing Survey Annual Schools Census Annual School-leavers and Associated Surveys Further Education Statistical Record		Register of Buildings of Special Architectural or Historical Interest Gypsy Sites Derelict Land Grant Applications/Payments Survey of Derelict Land in England in 1982 Survey of Mineral Workings in England 1982 Records of Mineral Working and Related Sites Geological Results from Planning Research Projects
DEPARTMENT OF EMPLOYMENT	Joint Unemployment Vacancies Operating System (JUVOS) National On-line Manpower Information System (NOMIS) Vacancy Circulation and Statistics Service (VACS) Employment Statistics System Database Census of Employment Local Employee Estimates Family Expenditure Survey New Earnings Survey		National Geological Data Bank Research Project Land Use Change Statistics Planning Appeals Information System Register of Waste Disposal Sites Monitoring Landscape Change Harmonised Monitoring of River Water Quality Lead Concentration in Drinking Water Nitrate Concentrations in Water Sources and Supplies Fluoridation of Water Supplies River Quality Survey 1980
DEPARTMENT OF ENERGY	Location of Boreholes/Wells Survey Line Display Petroleum Industry National Grid Area Reporting System (PINGAR)		River Quality Survey 1985 River Pollution Survey 1975 Pollution: Register of Discharges to land and rivers Contaminated Land Sites

DEPARTMENT OF THE ENVIRONMENT	Digitised Boundaries:
	  England and Wales Coastline and Standard Regions
	  1981 Districts and Wards
	  Planning Policy Areas
	  Rural Development Areas
	  Urban Areas of England
	Dwelling and Household data for General Improvement Areas and Housing Action Areas
	Home Improvement Grants
	English House Condition Survey 1981
	Housing Project Control System
			Monitored Domestic Energy Use Archive
			Radioactive Substances 1960 Act Record System
			Noise Measurements 1975-79
			Enterprise Zones Database
			Construction Industry − Output Data
			1981 Census of Population − Small Area Statistics

	Special Workplace Statistics
	1981 Census of Employment – Database
	Unemployment Database
	Central Postcode Directory
DEPARTMENT OF THE ENVIRONMENT: PROPERTY SERVICES AGENCY	Civil Estates Record Information Land and Hydrographic Surveys Location of Underground Services

DEPARTMENT OF HEALTH AND SOCIAL SECURITY	Social Security Statistics
	  Housing Benefit
	  National Insurance
	  Unemployment Benefit
	  Pensions
	  Supplementary Benefit
	Mental Health Enquiry
	Abortion Statistics
	NHS Property Estates Records

DEPARTMENT OF TRADE AND INDUSTRY	Sponsorship Assistance Management Information System Regional Data System
DTI BUSINESS STATISTICS OFFICE	Annual Census of Production Business Monitor Series
DTI WARREN SPRING LABORATORY	Spatial Distribution of Pollutant Emissions over the United Kingdom Ambient Air Quality Data Precipitation Composition
DEPARTMENT OF TRANSPORT	Road Network Information System Road Accidents Distribution Censuses of Traffic Rural Travel Survey 1984 Highway Maintenance Assessment Surveys Vehicle Keeper Register Drivers Licences HM Coastguard Incident Action Report Statistics
FORESTRY COMMISSION	Census of Woodland within Great Britain Forestry Sub-Compartment Database
GENERAL REGISTER OFFICE SCOTLAND	Old Parish Records Registers of Births, Deaths, Marriages etc. Census of Population National Health Service Central Register Central Postcode Directory
HOME OFFICE	Police National Computer Database Incidence of Fires in the United Kingdom
INLAND REVENUE	Wages and Incomes Database
INLAND REVEUE: VALUATION OFFICE	Rating Records Database Revenue and Compensation Database (including property transactions in England and Wales) Mineral Records Database
HM CUSTOMS AND EXCISE	VAT Records Database
HM LAND REGISTRY	The Register The Filed Plan The Public Index Map
METEOROLOGICAL OFFICE: MINISTRY OF DEFENCE	Weather Reports: Historical Computer Archives, climate, rainfall, anemograph, solar radiation, weather satellite datasets
DIRECTORATE OF MILITARY SURVEY: MINISTRY OF DEFENCE	Operational Navigation Charts Tactical Pilotage Charts Joint Operations Graphics Digital Terrain Elevation Data Digital Features Analysis-Data Production of Automated Charts of Europe (PACE)
HYDROGRAPHIC DEPARTMENT: MINISTRY OF DEFENCE	Admiralty Charts Admiralty List of Radio Signals Admiralty List of Lights and Fog Signals Admiralty Tide Tables Admiralty Sailing Directions Tidal Charts and Atlases Fisheries Charts
NATIONAL REMOTE SENSING CENTRE: BRITISH NATIONAL SPACE CENTRE	Landsat Imagery SPOT Imagery Archive of Geometrically Corrected Satellite Imagery Seasat, NIMBUS, NOAA Data
MINISTRY OF AGRICULTURE, FISHERIES AND FOOD	Annual Agriculture and Horticulture Census Underdrainage Statistics Agricultural Land – Classification Maps Less Favoured Areas Boundaries Internal Drainage District Boundaries Water Authority Boundaries Main Rivers Agroclimatic Datasets

NORTHERN IRELAND

Department of the Environment (Northern Ireland)
Land Registry
Statutory Charges Registry
Water Services Records
Housing Statistics
Road Service Records
Driver Licensing
Vehicle Registration and Licensing
Rateable Values/Properties
Schedule of Public Roads
OS(NI) Map products at 1:1250, 1:2500, 1:10,000 and 1:50,000 scales
Environment Protection
Planning Applications Records

Department of Health and Social Security (Northern Ireland)
Census of Population and Register Office Records

Department of Agriculture (Northern Ireland)
Agricultural Census, Soil Fertility Data, Animal Health Statistics

Department of Economic Development (Northern Ireland)
Computerised Unemployment System (COMPUS)

Valuation and Lands Office (Northern Ireland)
Property Database

OFFICE OF POPULATION CENSUSES AND SURVEYS
Census of Population
Medical Statistics
Vital Statistics
General Household Survey
Labour Force Survey
Central Postcode Directory
NHS Central Register

ORDNANCE SURVEY
Basic Scale Maps: 1:1250, 1:2500, and 1:10,000
Derived Maps: 1:10,000, 1:25,000
Small Scale Maps: 1:50,000, 1:250,000, 1:625,000
Digital Map Data: mainly 1:1250 and 1:2500
Digital Terrain Models at 1:50,000

REGISTERS OF SCOTLAND
General Register of Sasines
Land Register Database

SCOTTISH OFFICE

Department of Agriculture and Fisheries for Scotland
Agricultural Datasets
Annual Agricultural Census

Industry Department for Scotland
Regional Data System
Regional Assistance Scheme
Better Business Services
Factories Record
European Regional Development Fund Records
Census of Employment

Scottish Home and Health Department
NHS − Property Register
Scottish Health Services Costs

Scottish Development Department
Scottish area population statistics: digitised boundaries
Listed Buildings (Scotland)
Ancient monuments
Local Plans database
Planning Appeals
Industrial Site Register
Housing land annual return
River Pollution survey
Water Resources and Usage
Road Accident Statistics
Trunk Road database

WELSH OFFICE

Agriculture
Physical Information
Farm accounts

Demographic
Population, Births, Deaths

Economics and Industry
Labour Market: Advance Factories, Enterprise Zones, Development Board for Rural Wales − Factories, Ports traffic, Opening and Closure of Manufacturing Establishments

Transport
Railway Stations and Lines
Port traffic by goods and ships, Aircraft movements and airport accounts, road accidents

Education
Schools (Nursery, Primary Secondary) Welsh speaking (School leavers, handicapped pupils, higher & further education). Teacher Training, University of Wales

Housing
Tenure, New town construction, Demolitions and closures

Health
Mental Health, Mental Illness, Hospital Units, Discharges, Mental Handicap Hospitals and Units
General Health − Hospitals (Accommodation, Labour, Patients), Family Practitioners, Hospital Costings, Residential Accommodation for Elderly and Disabled.

Health Service Management Indicators, Bed Use Statistics, Hospital Activity.

Specialised Health Service Management Indicators, Waiting Lists, Cancer.

Environment Parks and Coasts, Tourists, Visitors, Derelict Land, Water Quality, Weather, Air Pollution.

Miscellaneous Cartographic − Ward and Community Boundaries

Planning Appeals, Planning Constraints, Register of Reservoirs, Coastal Survey, Air Photography, Thematic Maps, Topographic Maps

Planning Applications, Progress on Local Plans

Classified areas of Landscape, Register of Industrial Sites, Index of Deprivation, Survey of Contaminated Land.

Urban Programme Applications, Grant aided schemes from the European Regional Development Fund.

Major data sets of spatially referenced information typically held by local authorities

GENERAL Ordnance Survey Maps and Digital Data

Aerial Photographs

POPULATION Census of Population

Births and Deaths

Migration Data (from National Health Service)

Small Area Population Forecasts

Electoral Register

ECONOMIC Census of Employment

Small Area Unemployment Data

Vacancies

Business Directory

Tourism Directory

Census of Agriculture

Shopping Floorspace

LAND USE Land Use Surveys

Planning Applications

Housing Land Availability

Employment Land Availability

Derelict Land Sites

Waste Disposal Sites

Sites of Mineral Workings

Vacant Land Sites

Planning Policies and Proposals

Geological and Mining Records

ENVIRONMENTAL List of Buildings of Architectural or Historical Interest

Archaeological Sites and Monuments

Sites of Scientific Interest

Sites of Ecological Importance

Conservation Areas

Major Hazards

Atmospheric Pollution

Noise

Tree Preservation Orders

Amenity Open Space

HIGHWAYS AND TRANSPORTATION Highway Records

Volume and Type of Traffic

Bus Routes

Bridges

Street Lighting and Furniture

Sewer Records

Public Utilities Street Works

Road Traffic Orders

Road Accident Records

Site Investigation Records

Road Maintenance

Traffic Signals, Crossings, etc

Footpath Record

Streets Register

PROPERTY Comprehensive Property Index

Council Land Records

Council Property Records

Council Housing Stock

Building Regulations

Demolitions/Compulsory Purchase Orders

Dangerous, Defective and Neglected Buildings

Structural Surveys

Rating File

Housing Rents

Poor Housing

House Prices

CLIENT RECORDS Housing Rent Roll

School Rolls

College Rolls

Social Service Client Records

Location of Students

Borrowers of Library Books

Location of Incidents, including 999 calls

List of Trading Premises

Building Inspectors Records

Environmental Health Records

Free School Meals

ADMINISTRATIVE Statutory Boundaries

Policy Areas

Consultation Zones

Catchment Areas

Safeguarding Areas

Mailing List

Manpower Records

Voluntary Organisations

Amenity Societies

147

Appendix 4
Geographic information systems: functions and equipment

1 Geographic Information Systems, usually abbreviated to GIS, is a term which first appeared in general use in the late 1960s. It is normally used to describe computer facilities, which are used to handle data referenced to the spatial domain, with the capability to inter-relate datasets, to carry out functions to assist in their analysis and the presentation of the results.

GIS functions

2 Functions which are required in a comprehensive Geographic Information System have been summarised by Rhind (Reference 1) as:

Capabilities to retrieve data for any geographical area(s) specified by name or co-ordinates and on any combination of attributes.

Internal decision rules to assess which operations are feasible given the data available to the system.

Facilities to transform data from one co-ordinate system to another on an empirical basis using control points and to transform data on the basis of global expressions (eg. from co-ordinates in latitude and longitude to plotting co-ordinates on a specified map projection).

Facilities to update records for given geographical entities automatically by matching new and old records on the basis of a given, unique identifier and creating file linkages.

Facilities to convert data in string notation into raster notation and vice versa.

Facilities to build polygons from supplied line segments. The capability is to be possible where left hand and right hand area attributes are present and where no information other than segment end points is present.

Facilities to aggregate geographical entities, their geometric representation and their attributes to larger units, on the basis of some in-built hierarchy and on the basis of an externally supplied list, irrespective of the measurement scales of attributes.

Facilities to disaggregate geographical entities and their attributes, on the basis of the overlay of other entities and specified rules for the allocation of attributes over space.

Facilities either to carry out simple statistical sampling and summation or to re-format data appropriately for a standard statistical or modelling package.

Clearly documented interfaces to allow the transfer of data to or from the system at any stage.

Capabilities to generate tabulated records, graphs, histograms etc. and to map the data at scales and with symbols specified by the user.

Capability to monitor operations and to build up a composite picture of the user so that default values for parameters are progressively and automatically adjusted closer to the appropriate values with increasing use of the system.

In addition to these functions, interactive access to the system with a graphic computer terminal is pre-supposed.

Components of a GIS

3 The use of a Geographic Information System with the functions outlined above involves a chain of activities from the observation and collection of data through to its analysis and use in some decision-making process. A GIS may therefore be considered to have six main sub-systems:

    data acquisition and input
    data storage
    data retrieval
    data manipulation and analysis
    output
    systems management

4 *Data acquisition and input* involves the acquisition of data in both digital and analogue form, map data for processing, digitising and editing and the transformation of these data into a standardised pre-specified format.

5 *Data storage* involves data formats, hardware configurations, storage media and storage structure and is closely related to the data input and to the data retrieval sub-systems.

6 *Data retrieval* is a critical component since it directly affects the user's ability to get at the information behind the data and to structure the information to solve a specific problem. Different forms of data are retrievable in different ways. Map data are structured around the cartographic objects (points, lines or areas) often configured into networks and polygons. Non-map data may consists of values for attributes (variables) ordered by entity (observation or record). Database management systems are used to manage both types of data and are usually based on a data model, the standard models being the hierarchical, the network and the relational.

7 *Data manipulation and analysis* operations are frequently embedded into a GIS, although additional capabilities can be added as modules as required. Simple analyses might involve polygon overlay, conversion to different sets of polygons, tabulation of attributes by districts, density and area tabulations and the computation of attribute means. Spatial analyses may consist of terrain analyses, trend-surface analyses, spectral analyses, network solutions, location-allocation optimisation, model testing and simulation.

8 *Ouput* of information involves similar technology to that of automated cartography. Output may be by means of an

interactive display where the user can re-specify the image on demand or by the more traditional hard copy permanent imaging devices. Commonly used devices are screen copiers, electrostatic plotters, matrix printers, pen plotters, laser plotters and ink-jet plotters. A GIS may be able to output grid cell maps, network maps, flow maps, base reference maps, isopleth maps and perspective views. In addition statistical output can be displayed in the form of histograms, pie-charts and graphs.

9 *System management* should address the requirements for evaluation, and the organisational aspects of the use and effectiveness of the information transferred through the system. This includes the preservation of accuracy and reliability of data in order to avoid the production of misleading results.

Reference 1: D W Rhind, Geographical Information Systems in Britain, Quantitative Geography, edited by N Wrigley and R J Bennett, Routledge, 1981.

Appendix 5
Land and property databases

1 This note reviews the various systems used to handle land and property information. The main applications lie within local government and the national bodies with statutory responsibilites for holding property-related records. In general the land and property databases are inventory systems rather than Geographic Information Systems with capabilities for a wide range of data manipulations and analyses.

National authorities
2 The HM Land Registry is responsible for the registration of title to land and property in England and Wales to provide a simple, cost effective means of transfer of ownership.

3 Currently there are about 9.5 million registered titles out of an estimated total of 20 million. The number of titles is increasing at the rate of ½ million per year. The property register, identified by a title number, indicates the geographical location and extent of the registered property. It contains a written description and reference to an official plan (the filed plan) which is prepared for each title. In the case of a leasehold title it gives details of the lease. The proprietorship register gives the name and address of the owner; the charges register contains details of registered mortgages and notices of rights and interests relating to the property. The filed plan contains an outline of the boundary of the relevant property on a large-scale Ordnance Survey (OS) map. The 'public index map' is used to show whether any particular piece of land is registered or not. It is normally based on the largest published scale OS map of the area.

4 At present areas containing about 73% of the population are covered by compulsory registration and it is planned to extend these areas to cover 85% of the population by 1987.

5 The issue of information from the register, about a property on the register is restricted to the registered proprietor or persons holding his written authorisation. The Law Commission in its second report on Land Registration in 1985 has recommended, however, that the right of access to the register should be extended to everyone as is the case in Scotland and Northern Ireland. There is no confidentiality restriction on the public index map.

6 The HM Land Registry holds its information in the form of standard paper documents but is proceeding towards a computer-based system on two fronts. One involves the computerisation of the register in Plymouth District Office in the latter half of 1986, followed by a second district land registry office in 1987. The complete implementation in all 13 districts will be carried out by 1992. From the date of implementation in each district land registry all newly created entries will be held on the computer. The remaining entries will be converted in due course.

7 The other computerisation activity involves the use of OS digital map data as the basis for holding the official filed plan and public index map on computer. This procedure is now being implemented in a two-year pilot project at the Peterborough District Land Registry. The project will allow for the possibility of providing on-line access to map records by external users in the longer term.

8 The Valuation Office maintains Valuation Lists showing the addressess, descriptions and rating assessments of all rateable properties in England and Wales. Currently there are some 23 million entries (19 m domestic, 4 m non domestic). Rating survey details are held for the majority of properties, these include floorspace layouts and facilities. Valuation of all classes of property is also carried out for the purposes of capital taxation, eg capital gains tax, capital transfer tax, stamp duty etc. Data on the majority of transactions in property are also held; details include the nature of the interest transferred, the price or rent and the rights and restrictions relating to the property.

9 Work is also undertaken for other Government Departments and for the majority of local authorities, the Housing Corporation, regional health authorities and regional water authorities mainly relating to the acquisition and disposal of freehold and leasehold interests in land. The Valuation Office also maintains a mineral and mining records database to provide information on the risk of damage by mining subsidence and for valuation and planning control purposes.

10 The sources of Valuation Office information and the reasons for which it is obtained result in restrictions on its availablity to other users. In general, information held by the Valuation Office cannot be divulged to a third party without aggregation or the consent of the taxpayer, ratepayer or a party to the transaction. The Valuation Office has pointed out that there may be scope for amendments to the present rules governing disclosure of aggregated information which conceals the identity and details of individual properties.

11 The Valuation Office proposes to introduce micro-computers to support the preparation of the Valuation List, the recording of property transactions and the control of non-rating casework. The scope of the Valuation Office's database will need to await the Government's decision on the future of the rating system. If it is decided to abandon rates on domestic properties, as envisaged in the recent Green Paper on Local Government Finance, the coverage of the database would be reduced. The micro-computer based system being introduced to support the revaluation of non-domestic properties (now scheduled for 1990) could be expanded to deal with transactions and casework for domestic properties, although comprehensive coverage would not be required if rating were abandoned.

12 The Chief Valuer, Scotland, maintains valuation lists and

other relevant information relating to properties in Scotland, however, property transaction information is not restricted by confidentiality constraints as is the case in England and Wales.

13   The Land Register of Scotland holds the Sasine register of deeds as evidence of title to land in Scotland. This register contains records of present and past owners of 1¼ m properties in Scotland and includes details of the price paid and charges and burdens. The Sasine Register is gradually being replaced by the Land Register. At present the Land Register contains information on some 50,000 properties. The information on the legal title is computerised, but the map part of the Land Register still depends on OS paper maps. The registers are open to all.

14   The Valuation and Lands Office of Northern Ireland carries out similar functions to the Valuation Office and its database holds information relating to over 60,000 properties in Northern Ireland. The information includes current property description, occupation, ownership and tenure in addition to assessment and valuations. Information relating to individual properties is confidential.

15   The Land Registry of Northern Ireland maintains an open register giving details of ownership and other details relating to properties in Northern Ieland. The registry maps indicate the relevant locations and boundaries.

Local Government

16   Local authorities are among the main land and property owners within the United Kingdom and all authorities maintain some form of property based information system to support their property management activities. They range from simple property files to all-embracing databases. Such systems are characterised by holding discrete records of all streets and properties and most land parcels within a particular area.

17   The second report of the Conveyancing Committee (the Ferrand Committee) recommended in 1985 that local authorities should computerise their land and property information necessary to respond quickly to requests for local searches and additional enquiries. The Committee recognised that, as the computerisation of the HM Land Registry and local authorities' registers and records proceeded, there would be advantages in establishing computer to computer links to enable access from any single point to all other registers and records throughout the country. In the longer term such a process could lead to comprehensive property databases using unique property reference numbers and digitised maps.

Local Authority Management Information System (LAMIS)

18   LAMIS was developed in the early 1970's by ICL and implemented at Leeds. ICL moved on to develop the LAMIS approach on its more advanced 2900 range of computers in 1977/78 for Dudley Metropolitan Borough Council. This first implementation of LAMIS 2900 on-line property system was distinguishable from previous versions by the use of relational databases rather than serial files for storing data.

19   The system contains approximately 150,000 Unique Property Reference Numbers based on rating hereditaments in urban areas and on physical ground features in rural areas. These provide the common reference to link together information on a large number of files. The post code and land use codes were added to allow rapid retrievals using these key codes. Typical applications included the input to a register of

vacant land and the compilation of a directory of commercial establishments in various land uses.

20   To maximise the spatial search capabilities the LAMIS system allows for full boundary referencing of properties and land parcels. In a special project with the Ordnance Survey the area was selected for digital mapping with the resultant map data being specially restructured. The input to LAMIS of OS property boundary data in digital form allows the linkage, by spatial comparison, of application files such as planning applications and the property terrier. The addition of data on land and plant owned by the public utilities will add a further dimension to the database.

21   A number of other local authorities have implemented or are implementing LAMIS and LAMIS-type systems, notably Tameside, Hammersmith, Brent, Ealing, Doncaster, Birmingham and South Oxfordshire.

22   The original development of LAMIS in the early 1970's was intended to satisfy the requirements of local land changes by defining all property boundaries in spatial terms and producing an integrated set of data files serving the corporate needs of the Council. The main uses of LAMIS and LAMIS-type systems are to support:

(i)   the Planning Applications System: the requirements here are for an efficient system to deal with the large amounts of information resulting from the processing of applications for planning permission. The database approach enables linkages to be established to building control information and land availability statistics to identify land available for development.

(ii)   Land Potential and Development Monitoring: the uses include structure plan reviews, background for planning appeals, monitoring of Industrial Improvement Areas, completion of government returns, change information for the emergency services and the identification of new building starts for the Ordnance Survey.

(iii)   Housing Information Service: the information service relates to the dwelling stock managed by the local authority including structural characteristics, maintenance and renewal work. Information on tenants, housing waiting lists and related financial information on rents, arrears, payments to contractors, etc is also linked to the property file.

Gazetteer systems

23   The principal difference between gazetteer systems and LAMIS is that the central indexing is based on postal addresses not on spatially defined areas. Each property record has a single locational reference in the form of a grid reference of the property centroid, rather than digitised boundaries.

24   The Tyne and Wear Area Joint Information Service (JIS) has been operational since the late 1970's. It grew out of the National Gazetteer Pilot Study Project which was completed in 1975. The technical concept of the JIS system is simpler than that of LAMIS. The basis of the system is a street-map index together with a property gazetteer relating Unique Property Reference Numbers, centroid grid references, postal addresses, trading names, land use codes and indicators for enumeration districts and wards.

25 The JIS system has the advantages of simplicity and low updating costs; the level of usage is high not simply by planners but also by those in other local authority departments. These users predominantly receive address lists or statistical analyses, reflecting the fact that JIS provides routine processing and is not an on-line property system.

26 The main uses of the JIS system are the following:

  (i) Statistical analyses of catchment areas for services, dwelling counts and housing stocks;

  (ii) Forcasting and monitoring development activities, housing stock changes and land use changes;

  (iii) Identification of land available for housing or industry with records of every site approved for development;

  (iv) Provision of a Business Register detailing the products and services of firms.

JIS is also used routinely for locating streets, firms and properties for contacts, surveys and record keeping. The information is kept continuously up to date from departmental records, committee reports, statutory controls and site surveys. The system contains over 600,000 point referenced records of individual properties in the County.

Network based systems

27 A different approach to property referencing has been adopted by some local authorities such as Merseyside. The Merseyside Address Referencing System (MARS) is based on a system developed orginally by the Transport and Road Research Laboratory to provide a common locational referencing system for highway data, the Transport Referencing and Mapping System (TRAMS).

28 MARS is essentially a computerised directory of streets and addresses based on a digitised version of the County's road network. It uses segments of the road network as the basic structure; these are accurately positioned by recording the OS grid coordinates of the two end nodes to one metre resolution. Properties can be associated with segments of the road network since all properties have access to the road network and all postal addresses combine their address number or name with the street name.

29 Address gazetteers are an important product of MARS and represent the main requirement of its primary sponsor, the Police, who use it for their Command and Control System. The gazetteer facility can provide comprehensive lists of addresses for particular areas. A Street Index containing records for all potential incident locations is used for operational control and subsequent analysis of incident patterns.

30 MARS has played a central role in defining priority areas by enabling different sets of deprivation criteria to be tested in map form. Combinations of Census variables can be used to identify spatial concentrations of deprivation for comparison with existing policy and programme areas. Other MARS work in the land policy field has been concerned with the scale and pattern of derelict, neglected and vacant land using output from the Land Resources Monitoring System.

Partial property systems

31 Many local authorities have established property systems for particular sub-areas or datasets, often with in-house software. A notable example of partial property systems is the Central London Land Use System and Employment Register, CLUSTER, maintained by 23 London boroughs. CLUSTER has an updating system based on periodic resurvey (unlike the JIS and LAMIS continuous update approach) and a basic spatial unit corresponding to the land use parcel or overall building rather than the rating hereditament. This larger size of spatial unit is useful in the planning context of central London but does not equate to the JIS or LAMIS systems which contain records for every individual property.

32 The Land and Property databases are used in developing structure plans, local planning policies and for administrative purposes such as rate collection. The availablity of land (and its monitoring) for industrial and commercial purposes is also important to private developers. Typical additional uses of the systems include providing background for planning appeals, monitoring the effectiveness of Industrial Improvement Areas and Enterprise Zones, completing Government returns and providing information on changes and new building starts to the OS.

33 A further system which brings together elements from many other applications files is land charge searches whereby local enquiries are made into highway development, environmental health and planning in relation to particular properties. LAMSAC has considered the possibility of using computers to process local land charge searches and related activities, and many local authorities are now considering the introduction of computerised systems for this purpose. The link between property details and plot boundaries is of vital importance to this application.

34 The rating valuation file is a key source of basic information in cases where no central database has been established. It comprises the list of addresses for which rating assessments are made, a description of property, broad indications of usage, name and address of owner or agent. Property Gazetteers are maintained by a number of urban authorities to support not only the rating function but also the wider objectives of managing the housing stock of an area.

Appendix 6
Review of North American experience of current and potential uses of Geographic Information Systems

Prepared for
Committee of Enquiry into the Handling of
Geographic Information
Department of the Environment
by Tomlinson Associates Ltd, 17 Kippewa Drive,
Ottawa, Ontario, Canada K1S 3G3
August 1986

Introduction

At the outset it is appropriate to establish the terminology that is used in this report and to make a distinction between geographic information systems and such related activities as remote sensing and automated cartography. The lack of understanding of the differences, particularly between geographic information systems and automatic cartography, has been the source of much difficulty in the past twenty years in North America.

Data are considered to be 'spatial data' if they are individually or collectively referenced to location. Either specific location identifiers or nominal location identifiers are used to indicate positions in space related to other data. 'Specific location identifiers' allow spatial entities to be related either to some coordinate system or to other data elements by one or more spatial languages. Four types of locational primitive are used: points as abstractions of small phenomena or as surrogates for larger phenomena; line segments for linear features; arbitrary regular areas such as grid cells or pixels and irregular polygons for surface description.

'Nominal location identifiers' are typically names or code numbers such as postal codes, street addresses or arbitrary identification numbers which allow grouping and occasionally linkage of spatial entities but do not prescribe the position of the entities in space unless cross reference is made to another data set such as a map, which does specifically define the location of each name or code number involved.

Spatial data is a very broad term that can be used legitimately to refer to data sets as diverse as astronomical observations, topographic maps, the distribution of stress in a structure or the arrangement of chemical constituents in a sheet of photographic film.

'Geographical data' are herein regarded as those spatial data that result from observations or measurements of earth phenomena, particularly those data that describe natural resources, both renewable and non-renewable, as well as cultural and human resources. The spatial distribution of such phenomena is frequently shown graphically on maps, charts or images. Alternatively, the data may be in the form of lists of variables that are grouped by location and related places shown on maps. Typically the data describe such things as topography, soils, vegetation, water, surface geology, land use, administrative boundaries, transportation routes, political subdivisions, postal zones, facilities locations, lease boundaries, land parcel boundaries, land ownership, census districts, population distribution, income distribution, incidence of disease, land values, and so forth.

The spatial relationships between phenomena such as proximity, distance, connectedness, and the attributes of relationships between them such as flows, may be made explicit in a data set or may to a greater or lesser degree be implicit and calculable from the data record format and data structure. When the term spatial data is used in this report it generally refers to geographical data and encompasses the relationships that may be defined between them, although the capabilities of spatial analysis inherent in a modern geographic information system could be applied to other fields.

'Data gathering' encompasses the techniques of observation and measurement of phenomena. The various methods include remote sensing, survey, census and administrative record keeping. The primary purpose of data gathering is (or should be) the building of stores of data in forms that are amenable to subsequent use and which are relevant to the decisions that have to be made. Data gathering is a huge field in itself with important considerations that include criteria for selecting data to be gathered, the amount or lack of existing data, the cost of acquisition, gathering constraints, and factors of reliability, relevance to decisions, level of detail, rate of decay, institutional accessibility/confidentiality and so forth. 'Data handling, manipulation and analysis,' on the other hand, create the necessary bridge between data gathering and decision making. They comprise the organization, manipulation and display of data to allow the observed phenomena to be understood and, on the basis of that understanding and with a greater or lesser degree of accuracy, predicted. The lines between data handling, manipulation and analysis and the processes of data gathering and decision making are not hard and fast. There is often a degree of overlap. Data handling can occur during the gathering of data; remotely sensed images can be processed, survey results can be reduced, census can be summarized, and records can be reformatted in a variety of ways. Similarly, the path between data gathering and decision making can be very short; a single fact, quickly acquired, may influence a decision. In general terms, however, data handling, manipulation and analysis make up the step that transforms data into useful information from which decisions may be taken. 'Decision making' processes at administrative, management or policy levels, can then result.

The term 'system' or 'digital system' as used in this report describes a combination of manual and computer-aided processes for data handling involving computer hardware, software and special-purpose computer equipment.

An 'information system' is a system that also has the ability to handle, manipulate and analyze data to provide information that is useful in the decision-making processes of an individual or an organization. We define a Geographic Information System (GIS) as a digital system for the analysis and manipulation of a full range of geographical data, with associated subsystems for digitizing and other forms of input and for cartography and other forms of display used in the context of decision making. The emphasis is clearly on the analysis and manipulation functions, and in the GIS field they provide the primary motivation for using digital methods; if the intention were not to analyze or manipulate, there would be no point in converting the geographical data to digital form in the first place.

Automated cartography on the other hand is defined as the use of computer-based systems for the more efficient production of maps; such systems may replace various forms of manual activity associated with map production, such as scaling, editing, colour separation, symbolization or typesetting. The systems which have been developed for automated cartography use different data structures and offer a quite different set of functions from those common in geographic information systems, and in general the two types of systems are not highly compatible.

A futher general point should be made in introducing the discussion; we are concerned as much with the institutional and human framework of GIS use and development as with the technical framework. Although this point will be repeated at several levels, it is worth stressing at the outset that in dealing with a relatively new technology such as GIS we have found over and over again in North America that the technical problems are minor in comparison with the human ones. The success or failure of a GIS effort has rarely depended on technical factors, and almost always on institutional or managerial ones.

1. Application areas
The future of GIS will depend to a large extent on the degree to which the needs of various application areas can be integrated and met by one type of product. The enormous growth in GIS activity over the last few years has been led by one or two application fields, and there have been distinct differences in the form development has taken, and the meaning attached to the term GIS itself, in each area. The strategy we will take in this report will therefore be first to review GIS in the context of each area, and then to take a more general perspective.

There are of course significant differences between the North American and UK contexts, and the ways in which application areas have developed, so the following discussion uses a North American taxonomy. For example, forestry is very clearly a major application area with its own unique problems and concerns in North America, whereas it is likely to be subsumed under 'agriculture and countryside' in the UK. The sectoral divisions listed below are those which we believe have played an important role in the historical development and exploitation of GIS technology in North America.

1.1 FORESTRY
Forestry has been responsible for a significant growth in GIS use in the past five years. Ideally, GIS technology would be used for the updating and maintenance of a current forest inventory and for modelling and planning forest management activities such as cutting and silviculture, road construction, watershed conservation, etc. In other words the true advantages of GIS accrue only when emphasis is placed on the manipulation, analysis and modelling of spatial data in an information system; there is a broad acceptance of the principle that the use of computers for cartography alone is not cost effective. In practice GIS's have often been used for little more than automation of forest inventory cartography, because of limitations in the functionality of software or resistance to GIS approaches on the part of forest managers. This situation will change slowly. It is of course a general problem affecting other areas beside forestry.

In a typical North American forest management agency the primary cartographic tool for management is the forest inventory. It is prepared for each map sheet in the agency's area on a regular cycle, which requires flying and interpreting aerial photography, conducting operational traverses on the ground, and manual cartography. In one such agency with which we are familiar the number of sheets is 5,000, and the update cycle 20 years. Events which affect the inventory, such as fire, cutting and silviculture, are not added to the basic inventory.

The earliest motivation for GIS in forestry was the ability to update the inventory on a continuous basis by topological overlay of records of events, reducing the average age of the inventory from the existing 10 years to a few weeks. More sophisticated uses include calculation of cuttable timber, modelling fire, and supporting the planning of management decisions. To our knowledge every significant forest management agency in North America either has now installed a GIS, or is in some stage of GIS acquisition. We know of no agency which has rejected GIS in the past three years.

The number of currently installed systems in this sector in 1986 is estimated at 100, in federal, state and provincial regulatory and management agencies and in the private sector of the forest industry. This figure is based on a census of Canadian agencies, where we know of 13 at the present time, and an estimate of the relative sizes of the Canadian and US industries. Systems have been supplied by the private sector (Environmental Systems Research Institute [ESRI], Comarc, Intergraph) or developed internally (Map Overlay Statistical System [MOSS]) and there are also public domain systems with various levels of private sector support (MOSS, several raster-based systems). Note that many of these systems fall short of supporting a complete range of forms of spatial analysis on all types of locational primitives. Almost all major agencies and companies in the North American forest products industry would claim some level of involvement in GIS at the present time: the figure of 100 is an estimate of the number who currently maintain a significant establishment of hardware, software and personnel with recognizable GIS functionality.

Several forms of *de facto* standardization have emerged in various areas of the industry. In New Brunswick the private and public sectors coordinated efforts by acquiring identical systems – the New Brunswick Department of Natural Resources, Forest Management Branch, and the two major forest products companies have all acquired ESRI. The US Forest Service is in the process of determining a coordinated approach to GIS for its 8 regions and 155 forests.

The potential market in North America is approximately 500, on the assumption that all US National Forests will acquire a minimal system, together with the regional and national offices of the Forest Service, and that in large agencies such as the Forest Resources Inventory of the Ontario Ministry of Natural Resources there will be systems in each of the major regions. Barring major downturns in the industry we would expect this to be reached within the next 5 to 10 years. At the same time the emphasis will shift more and more to analytic rather than cartographic capabilities, so there will be a growing replacement of simple systems by those with better functionality. This will be an expensive change for some agencies who will find that their early systems or databases lack upward compatibility, often because of the lack of topology in the data structures.

Several general points can be drawn from the experience in North America to date. First, virtually all forest management agencies which have attempted to 'go it alone' and undertake development of systems, either in house or through contract, have met with disaster, in some cases repeatedly. The development of a GIS is a highly centralized function requiring resources far beyond those of a single forest management agency, and this applies to other applicaton areas as well. Second, the resources of a single agency similarly do not permit effective, informed evaluation of vendor products. If acquisition is uncoordinated and agencies are permitted to go their own way, the result is a diversity of incompatible systems which are in most cases inappropriate solutions to GIS needs. We have found tremendous variation both in the capabilities of vendor systems and in the needs of each agency, and matching these is a difficult and complex task. As the field matures the differences between systems will presumably lessen, but the variation in the nature of each agency's workload requirements will if anything increase in response to more and more sophisticated forms of analysis and modelling. Put another way, we have found that there is no such thing as a common denominator system for the forest industry: the natural tendency is for diversity in agency needs and vendor responses, rather than uniformity.

Part of the reason for this diversity lies in the mandates under which many responsible agencies must operate. The typical US National Forest has a mandate to manage not only the forests in its area, but also the wildlife, mineral and recreation resources. The purpose of management must also allow for the need to conserve as well as to extract. The typical Forest Service GIS will be used to manage road facilities, archaeological sites, wildlife habitat and a host of other geographical features. The relative emphasis on each of these varies tremendously from forest to forest; several, for example, have significant coal reserves and others have no trees.

There has been no attempt to coordinate GIS technology nationally in either the US or Canada in the forest industry. In Canada the Forestry Service of the Federal Government has an advisory mandate and its acquisition of an ESRI system has possibly had some influence on developments in provincial agencies and companies, along the lines of a centre of excellence, but this is far from certain. We are not aware of any overall study of sectoral needs in either country, or of any effort to promote or fund system development to meet those needs.

There are several factors accounting for the very rapid growth of GIS activity in the forest industry recently. First, effective forest management has been a significant societal concern and has attracted government funding. Second, as we have already argued, GIS technology is seen as an effective solution to the problem of maintaining a current resource inventory, since reports of recent burns, cutting and silviculture can be used to update a digital inventory immediately, resulting in an update cycle of a few months rather than years. Third, it is attractive as a decision tool to aid in scheduling cutting and other management activities. Finally, because of the multithematic nature of a GIS data base it is possible to provide simultaneous consideration to a number of issues in developing management plans. All of these factors, combined with the perception that GIS technology is affordable, have given vendors a very active market in the past five years or so in North America. Note that in the last two areas at least the functionality offered by the GIS is substantially new, and does not automate an existing manual process.

1.2 PROPERTY AND LAND PARCEL DATA

The acronyms LIS and LRIS (Land Information System and Land Related Information System respectively) are often used in this application area, reflecting the relative importance of survey data and emphasis on retrieval rather than analysis functions. Most major cities and some counties have some experience in building parcel systems, in many cases dating back to the earliest days of GIS, but in general state or national systems have not been considered in North America because land registration is usually a local responsibility. Fragmentation of urban local government is also a problem in the US and in some Canadian provinces. Nevertheless the special cadastral problems of the maritime provinces have led to investment there in an inter-provincial system (Land Registration Information Service [LRIS], using CARIS software by Universal Systems, Ltd.).

There is a potential market of some 1000 systems among the 500 major cities of 50,000+ population and 3000 counties of the US, and perhaps a tenth of that in Canada. The number of these agencies with significant GIS investment has not changed markedly in the past ten years, however. Although several cities established an early presence in the field, their experiences were often negative, and projects were frequently abandoned. More recently improved software and cheaper hardware has meant a greater rate of success, but we are still very low on the growth curve.

The functionality needed in an LIS is well short of full GIS capabilites. In many cases all that is needed is a geocoding of parcel records to allow spatial forms of retrieval: digitizing of parcel outlines is useful for cartographic applications. It is unlikely that many municipalities will advance to the stage of creating a full urban geographic information system by integrating transportation and utility data and developing applications in urban development and planning, at least in the next five to ten years. Cities which have tried to do this have discovered the difficulties of working across wide ranges of scales, from the 1:1,000 or greater of land parcel data to the 1:50,000 suitable for such applications as emergency facility planning, bus system scheduling and shopping centre development. There is very little suitable applications software in these areas at present. In summary, automated cartography and retrieval will probably remain the major concerns of land parcel systems in the next ten years, and confidentiality and local responsibility will remain barriers to large scale system integration in North America. These needs will likely continue to be met by vendors of automated cartography systems (Intergraph and Synercom), and Data Base Management Systems (DBMS). Furthermore it is unlikely that many municipalities will reach the stage of giving the digital data legal status as a cadastre: problems of accuracy and confidentiality are likely to prevent that for some time yet.

The influence of government in this applications area has largely been driven by concern for data quality, rather than technical development. A number of system acquisitions have been sponsored by senior governments for demonstration purposes, and as a means of promoting quality and format standards. However with a few exceptions in the very early days, there have been no major system developments undertaken within this field, most needs apparently being satisfied by hardware and software already available. A major experiment is currently under way in Ontario to apply ESRI software to the needs of municipal data bases for such varied applications as policing, emergency facilities, transit systems and schools. But it is too early to tell whether this will lead to widespread adoption of similar approaches in other cities and to a significant movement up the growth curve.

1.3   UTILITIES

Telephone, electric and gas utilities operate in both private and public sectors in North America, and are generally less centralized than in the UK. It is useful to distinguish between applications at large and small geographic scales. Large scale applications include monitoring of pipeline and cable layouts, locations of poles and transformers, and as in the case of land parcel systems, combine needs for cartography and spatial retrieval. Traditionally these have similarly been met by automatic cartography vendors, particularly Synercom, and by DBMS's. However it is possible that demand for more sophisticated forms of spatial retrieval and layout planning will lead to reorientation towards the GIS model in the long term.

Small scale applications include planning of facilities and transmission lines to minimize economic, social and environmental cost, and demand forecasting. Some utilities have built large data bases for such purposes, and also make use of digital spatial data from other sources. For example the location of microwave repeaters is a significant application for digital topographic data. Such applications are highly specialized and idiosyncratic, and have largely been handled by specialized software rather than generalized systems. In the long term as GIS software stabilizes and develops it is likely that the advantages of better input and output capabilities and easy exchange of data formats will make such systems an attractive option for these application area. We have not yet reached the beginning of the growth curve in this sector, but in the next five to ten years a significant proportion of this market will probably move away from existing automated cartography systems into GIS, and several vendors appear to have anticipated this trend already, as illustrated by Synercom's marketing of Odyssey. There are some 200 utilities in North America with potential interest in GIS applications.

1.4   TRANSPORT, FACILITY AND DISTRIBUTION PLANNING

In addition to both public and private sector transportation agencies, much of the work in this application area is carried out by research contractors, such as market research firms and universities. A 1982 paper (Raker, D.S., Computer mapping and geographic information systems for market research. *Proceedings, National Computer Graphics Association 2* 925-933) listed 26 companies providing either GIS services or geographical data, or both, in this area. Although market research frequently calls for sophisticated forms of spatial analysis, such as site selection and spatial interaction modelling, GIS vendors have not made any significant penetration into this market to date. Instead most companies rely on a combination of standard statistical packages (SPSS [Statistical Package for the Social Sciences], SAS [Statistical Analysis System]), Data Base Management Systems and thematic mapping packages developed in house or acquired from vendors (GIMMS [Geographic Information Manipulation and Mapping System]). Software and data sales are often a significant contribution to income.

There is every indication that this will be the major growth area for GIS applications in the next ten years. Software for these forms of spatial analysis is at present rudimentary but developing rapidly (for example ESRI's NETWORK). There is a pressing need for the ability to handle multiple geographical data formats, scales and feature types, and hierarchical aggregation of features, in processing socioeconomic data, in combination with advanced forms of spatial analysis and map display. In addition to market research firms, the potential market includes major retailers, school systems, transportation and distribution companies and agencies who need to solve routing and scheduling problems on networks, and the direct mail industry. It is not at all unreasonable to visualize a potential market of 1,000 systems in North America, or even ten times that, but to date there has been very little interaction between GIS vendors and those software house which have traditionally operated in this market.

1.5   CIVIL ENGINEERING

A major use of digital topographic data is in large scale civil engineering design, such as cut and fill operations for highway construction. The first digital developments in this field derived from the photogrammetric operation, which is the primary data source. More recently efforts have been made to add more sophisticated capabilities to photogrammetric systems, noteably by Wild (System 9) and Kern, and to interface them with automated cartographic systems and GIS's. Government agencies in North America (US Geological Survey, Canada Department of Energy, Mines and Resources) have been fairly active in this field in attempting to develop common data standards and interchange formats.

There are some 50 major systems installed in civil engineering contractors and government agencies in Canada (based on a personal census) and perhaps ten times that number in the US. This figure is unlikely to change in the next ten years as this is not a growth industry. However in both Canada and the US rapid growth is currently occurring in the significance of digital topographic data for defence, because of its role in a number of new weapons systems, including Cruise, and because of the general buildup of defence budgets in the industrialized world. This has drawn attention to the importance of data quality, and the need for sophisticated capabilities for topographic data editing as well as acquisition. These needs are presently being met by enchancements to automatic cartography systems (Intergraph) and it is not yet clear whether they will lead in the long term to any significant covergence with GIS's.

1.6   AGRICULTURE AND ENVIRONMENT

From a Canadian perspective these are the original application areas of GIS: projects in federal evironment and agriculture agencies, in the form of the Canada Geographic Information (CGIS) and Canadian Soil Information System (CanSIS) respectively, both date back over 20 years. The use of GIS approaches can be traced to the need to measure the area of land resources, the need to reclassify and dissolve prior to display, and the need to overlay and spatially compare data sets. These remain among the most basic justifications for GIS technology. Although both systems were developed largely inhouse in government agencies, the capabilites of CanSIS are now provided by a number of cost-effective vendor products. On the other hand CGIS remains to some extent unique: no vendor has yet developed a system with comparable capabilities in bulk data input and national data archiving.

The environment market is much less significant in most countries than it was fifteen years ago, except in specialized areas. GIS technology is of considerable interest in land management, particularly of national parks and other federal, state and provincial lands, and has been adopted in both the US and Canada: there is a potential of perhaps 10 systems in Canada and at least ten times that number in the federal land management agencies in the US. However in the present mood of fiscal restraint in both countries neither figure is likely to be reached in the immediate future.

There appears to be a rather greater support at the present time for global environmental monitoring and analysis programs, under such agencies as United Nations Environment Programme (UNEP) and Food Agriculture Organization (FAO). However although there will have to be massive investment in data gathering, it is unlikely that there will be a large market for systems.

In agriculture the main issue is time dependence. Although a large amount of research has been conducted on the interpretation of argicultural data from remotely sensed imagery, and there are no major technical problems in interfacing image processing systems with GIS's, there remain the conceptual problems of classification and interpretation. GIS designers have typically assumed accurate data: to date no significant progress has been made in designing systems to process and analyze uncertain data. Similarly the conventional GIS has no explicit means of handling data which is time-dependent or longitudinal, yet this is characteristic of data in a number of areas besides agriculture; marine environmental monitoring is a good example, as is climatology.

1.7 INHIBITING FACTORS

A number of factors have clearly inhibited the wider use of GIS technology in North America, and many will continue to do so for some time to come. First, there is no overall program of large scale digital base mapping, and as yet no suitable base is widely available. This is an inhibiting factor, but not in our estimation a major one. To be suitable, such a base would have to exist at a number of scales, since it would be too elaborate and expensive for users to derive specific scales by generalization from a largest common denominator scale. It would have to exist in a number of formats. And since it would be so large that no average user could expect to maintain a private copy, it would have to be available at very short notice, probably on line. Such base maps as exist, notably World Data Base (WDB) I and II and state and county outline files, have been widely distributed, and have perhaps tipped the balance for a few users in deciding to opt for GIS technology. However other factors in that balance seem much more important. And although WDB II is more detailed and comprehensive than WDB I, the former remains much more widely used as a world digital data base because it is much smaller and therefore more readily manipulated.

Data availability has been a significant factor in some parts of North America and in some application areas, but somewhat oddly seems more likely to work in reverse: it seems easier to justify data collection if the data are to go into a GIS, where they can more economically and quickly be manipulated and analyzed in concert with other data. Much of the justification for the North Slope Borough acquisition of data in Alaska centred on the way these data would be used in a GIS and lead to better control over a fragile ecosystem. Similar points could be made about UNEP. We suspect that there is very little user support in North America for the establishment of a large, general-purpose bank of digital cartographic data, because experience in other contexts and with national spatial data archives such as CGIS has taught that to date such projects simply cannot be made readily accessible with widely applicable, current data at reasonable cost.

Digitization is certainly an important factor of cost-effectiveness if taken seriously, since it will usually be the largest element of operating cost, although not typically in the case of marketing applications. Unfortunately the trend of its relative cost in the near term is likely to be for the worse, since hardware and software costs per unit of capability will certainly drop, whereas it is difficult to see how the costs of manual digitizing will change significantly. Certainly scanning will become cheaper, but document preparation will remain a significant cost element however sophisticated the vectorization software.

Software is a very important element. Although most of the scientific developments had taken place in the sixties. It took a further ten years before any one group could make sufficient investment in software development to produce a commercial, transportable software package, and of course none of the present systems approaches the ideal. The price of a package is determined by dividing the investment by the perceived market. It seems reasonable to assume that the amount of investment in future generations of GIS software will grow linearly, whereas the potential market will grow geometrically, leading to significant reductions in price. At present we have seem some hundreds of systems (notably Intergraph, Synercom and ESRI) installed at a software price of $100k, to the nearest order of magnitude, and the market is far from saturated at that price. It is entirely possible that the market would be ten times larger if the price were $10k.

It is tempting to think that much of the current boom in GIS's acquisitions stems from the convergence of increasing software sophistication and decreasing hardware costs, so that a turnkey system can now be installed for around $250k, based on VAX or Prime hardware. However many of the current systems would run perfectly well on timeshared mainframes. Our own view is that the typical purchaser agency would rather incur a one-time capital expense of $250k than a continuing and variable expense of perhaps $100k per year, because it offers a stronger position against an uncertain budgetary future. Furthermore, the turnkey system is a tangible item over which the system manager can exert substantial control.

Certainly we will continue to see reductions in the costs of GIS processors and peripherals, and probably also in mass storage. Software will therefore continue to increase in relative importance in overall startup costs.

157

User awareness is an extremely important factor, along with other behavioural and sociological issues. Information about GIS has tended to be disseminated through personal contacts, reports and meetings rather than through formal organizations, journals and textbooks, and there has been no systematic approach to GIS in the educational system. The term is encountered in courses in departments of geography, surveying and forestry, in other words primarily in application areas, by students with little or no technical background. There have been attempts in North America by various application areas to coordinate the development of specialist programs, but these will not be successful as long as they are perceived to benefit one application area over another. The federal governments are unlikely to have much influence on the educational system until they can coordinate their own interest in geographical data handling, and there is little sign of that at present. The greatest progress has been made at provincial government level in Canada, where steps have been taken to coordinate the development of university and college programs.

This has led to a characteristic pattern which we have observed in many agencies. Typically an individual with some background in computing will hear of or see a geographic information system or automatic cartography system in operation, and by attending conferences or workshops become something of an enthusiast and promoter within his own agency. With luck he will eventually assemble sufficient resources to acquire a system, and naturally be named manager of it. The rest of the agency will be very happy to know that the group are involved in the new technology, but equally happy that all obligation to understand rests with the resident expert.

We have found that the only effective way out of this sociological impasse is to undertake a complete and comprehensive functional requirements study with the full cooperation of the agency director. The ways in which GIS's products will be used by the agency must be documented well before the GIS is acquired and installed, so that they become obligations on the part of the agency staff to use and understand the function of the system.

The only possible alternative would be to ensure that all staff be introduced to GIS as part of their basic education, whatever their application area: if this were possible, it would take decades to achieve. In short, we believe that the greatest obstacle to greater GIS use will continue to be the human problem of introducing a new technology which requires not only a new way of doing things, but whose main purpose is to permit the agency to do a host of things which it has not done before, and in many cases does not understand.

Extensive use of digital approaches to geographical data handling draws attention to a number of technical issues which are often transparent or ignored in manual processing. It is necessary to be explicit about such formerly implicit issues as accuracy, precision and generalization. Coordinate systems, projections and transformations must be specified precisely, because there are no globally accepted standards. In North America we will continue to suffer through a complex maze of different coordinate systems; Universal Transverse Mercator and latitude/longitude at national scales, but a variety of systems at state and provincial levels. The UK with its smaller area and more centralized administration seems more likely to avoid these problems.

A number of developments in the next few years are likely to

affect the general level of user awareness of GIS. Although the first text in GIS has only recently appeared (by P.A. Burrough, published by Oxford University Press), the set of teaching materials is likely to grow rapidly in response to obvious need. New journals are appearing. Several universities have instituted programs at undergraduate and graduate levels, and there are technical diploma programs available at several institutions. The National Science Foundation is seeking reactions to a proposal for a National Centre for Geographic Information Analysis which could have a major coordinating role. But despite this activity, the supply of qualified PhD's in GIS to staff these new programs and carry out the research which will allow the field to develop remains woefully inadequate to meet current demand.

1.8 SPATIAL DATA SETS

Many government agencies have undertaken surveys of existing spatial data sets in order to improve access and reduce duplication. Notable examples are the USGS in the late 1970's, EMR in 1984, Ontario and Quebec in the early 1980's. In some cases these include only data sets where the locational identifier is specific, in the form of a coordinate pair: in others the lcoational identifier can be nominal; a pointer to a location specific file, as in the case of a street address. Some of these studies have provided directories only, but others have gone on to estimate existing and potential usage: the EMR study is an example. In the latter case the results were almost uniformly disappointing: where data sets had been compiled at least partly for use by others, the level of use had almost always been overestimated.

In general where such studies have been made at the federal level they have been undertaken by a single agency with little prospect of continuing update. On the other hand provincial governments such as Ontario have been more successful at undertaking coordinated, multi-agency surveys, typically by a Cabinet decision assigning responsibility to one Department, Natural Resources in the case of Ontario.

There seems little prospect of coordination of key spatial data sets in North America because of problems of confidentiality and division of responsibility between the three levels of governments. In Canada the administrative records from Revenue Canada have been linked to postcodes, and Statistics Canada has linked postcodes to census areas. But the detailed postcode boundary network is not planar and has not been digitized, and the lowest level of census areas changes in each census. Land parcel files are a municipal responsibility in most provinces and are linked to the system of municipal election boundaries, which do not respect any of the federal units. Because of these problems there has been a move to the concept of a municipal census in some cities: certainly the trend in North America at present is for a decrease in the power of central governments to coordinate socio-economic data gathering.

Several general points can be made based on knowledge of the development and used of digital spatial data sets, in both Canada and the US. First, the potential for exchange and common use of digital cartographic data in both counties is not yet being realized. Although the data structures in use in various systems are broadly compatible, there is little or no standardization in detail, and each of the major systems currently in use has developed independently to satisfy the specific needs of individual agencies. Problems of coordinate systems, scales and projections, lack of edgematching and incompatible use of tiling and framing remain major barriers to exchange and sharing, and encourage duplication of effort.

Efforts which have been made to standardize formats, such as the US Geological Survey Digital Line Graphics (DLG) and the Canadian Standard Data Transfer Format (SDTF), have had little effect, whereas there has been rather more *de facto* standardization because of the widespread installation of systems from the same vendor; Intergraph's Standard Interchange Format (SIF) has become quite successful for this reason.

The option of making digital data rather than buying it or copying if from another system will remain popular as more and more vendors enter the market with unique data formats. But although the incentive to go digital will grow as software and hardware costs drop and as functionality improves, this will not immediately affect the amount of data sharing, since it will reduce not only the cost of buying or importing data, but also the hardware and software entry costs related to digitizing it locally. So it is likely that duplication will remain common for some time to come. Sharing is also hindered by lack of information on what is available, and lack of rapid access, and no fundamental change is likely in either factor in the near future.

In summary, we would expect in principle that the ready availability of data on a wide range of geographical themes should be one of the major benefits of the trend toward digital cartography and GIS. In practice the picture in North America in 1986 is one of very incomplete coverage both regionally and thematically, due more to historical and behavioural factors than to design. Each major system has developed independently and has felt little pressure to standardize, or to establish interchange capabilities. As a result very little actual transfer of data takes place, although there is more frequent use of another agency's analytic facilities. At the same time digital capabilities are being acquired by all types of agencies at a rapid rate in the absence of coordination, and the popularity of a small selection of systems is creating a situation of informal and arbitrary standardization.

2   Costs and benefits

There has been very little in the way of formal accounting of GIS operations, largely because in most agencies they are still regarded as experimental. The only GIS with which we are familiar which has been in production for long enough with stable hardware and software to permit any kind of realistic accounting is CGIS, which has kept accurate and detailed records of operational costs for some time (see the paper by Goodchild and Rizzo in the Proceedings of the Seattle International Symposium on Spatial Data Handling, July 1986). Experience in the U.S. software engineering field indicates costs of such record keeping would run about 2-3% of total project costs.

A number of agencies have carried out some form of benchmark testing as part of their GIS acquisition process, although others have relied entirely on vendor claims, in some cases with disastrous results. However benchmark testing has almost always been designed to test the existence of functionality, rather than to test performance speed and thereby to project production cost. This merely reflects the immature state of the GIS software industry, and will presumably change slowly over time as more and more acquiring agencies use benchmark tests to determine the extent to which the tendered system will or will not perform the required GIS task workload within the prescribed schedule. Workload estimation will also become

increasingly important if a significant market develops for GIS service, as distinct from system sales.

The unit of output of a GIS is a processed information product, in the form of a map, table or list. To assess the cost of this product would require a complex set of rules for determining cost accrual: these are fairly straightforward in the case of the direct costs of the product, but less so for the capital costs of the system and the costs of inputting the data sets from which the product was derived. We are not aware of any system which has attempted to establish such a set of rules.

Far more difficult is the determination of the product's benefit. As an item of information its benefit can be defined by comparing the eventual outcome of decisions made using it, and decisions made without it. In almost all circumstances such outcomes would be uncertain in both cases, so probabilities would have to be estimated. Our discussion of this issue in the context of the US Forest Service supposed that it might be possible to identify a small number of decisions in which some of these problems could be resolved, but that it would be next to impossible to compute total benefit in any sensible way. We proposed therefore to institute a program of identifying and tracking suitable decisions, in order to form the basis for a largely qualitative evaluation of benefit.

In those cases where the GIS product replaces one which would otherwise be generated by hand, it is relatively easy to resolve the cost/benefit issue by comparing the costs of the two methods: the benefits are presumably the same. Where this has been done, as for example in the Ontario Forest Resource Inventory functional requirements study, the GIS option is usually cheaper by an order of magnitude. However this approach glosses over the fact that many of the products would not have been requested or generated if the GIS option were not available.

These problems are not unique to GIS, but occur in similar fashion in all information systems, and in many other data processing applications as well. It is clear that without a great deal more experience of production, the case for GIS's will not be made on the basis of a direct, quantitative evaluation of tangible benefits, but must include less direct and more subjective issues such as improved access to information, more objective and better informed decision-making, and improved data quality.

Similarly we are not aware of any direct attempt to compare the costs and benefits of an installed operating GIS with those estimated during the planning process. There is of course abundant anecdotal information of this kind, little of it of much substance. The practice of comprehensive planning for GIS acquisition has developed only in the past five years or so, and it is still relatively unusual to find that any attempt was made to estimate costs or benefits in advance. However it is likely that this sort of evaluation will be required for some of the pending major acquisitions, and this will provide an opportunity to conduct an objective evaluation for the first time.

3   Technical developments

3.1   DIGITIZING

Reference has already been made to the suspicion that future technical developments will not lead to any marked improvement in manual digitizing methods. There is however considerable room for improvement is automated procedures, and consequent reductions in data input costs. More sensitive scanners will allow greater discrimination between data and noise. Better software, with properties verging on artificial intelligence, will produce more accurate vectorization and feature identification. Rather than employing directed scanning, the systems of the future will probably scan the document in a simple raster, and rely on processing to track features logically rather than phyiscally, because memory is becoming less and less of a constraint. There are now good raster edit systems on the market, and future systems will optimize the relative use of raster and vector edit to minimize overall edit time and cost. Finally, better hardware and software will allow greater use of raw documents and less need for specialized document preparation and scribing. In the long term there is plently of scope for redesign of cartographic documents to make scanning easier and more accurate; use of special fluorescent inks and bar codes, for example.

These developments may mean that raster input will eventually replace vector digitizing for complex documents and bulk input. Manual digitizing will remain a low capital cost option however for input of simple documents and for systems requiring low input volume.

3.2   RASTER vs VECTOR

In most of the application areas listed above the current market is dominated by vector rather than raster systems. To clarify a little, since all systems are to some extent hybrids, the distinction refers to the form of storage used for the bulk of the data base, and to the data structure used in the majority of the system's data processing.

The primary concern in choosing between raster and vector systems is the nature of the data in the particular application area, and in some areas it has taken years of experiment to resolve this issue. In parcel systems the need to work with arbitrarily shaped but precisely located lot features has led to vector systems: the vector approach allows the definition of entities that are linked to statutory responsibility. The nature of elevation data argues for raster storage and most satellite-based imagery originates in that form. These conclusions are fundamental to the nature of each application area and are unlikely to change.

On the other hand any vendor wishing to market a comprehensive system whose applicability spans several application areas must consider both data types, and transfers between them. Several vector systems (ESRI, Intergraph) offer raster-based functionality for handling topographic data. Ultimately, all comprehensive GIS's will recognize four locational primitives, point, line, area and pixel, and will permit appropriate processes on all four. And there will continue to be systems designed primarily for image processing which retain the raster as the central mode of data storage but permit vector features for specific functions.

3.3   SOURCES OF RESEARCH AND DEVELOPMENT FUNDING

In North America at the present time no government agency has a general mandate to fund research and development in geographical data handling: those agencies which do so all have specialized needs in defined application areas, or a general mandate to fund basic research. There is no indication that this situation will change in the next five to ten years.

The state of funding thus depends very much on the state of each application area's responsible agencies. At the present time in North America public attention on environmental issues is perhaps less focussed than before and forestry, oil and gas are probably past their recent peaks. On the other hand substantial amounts of money are being allocated to research into water resources, and we can therefore expect some interest in GIS applications in this field. As noted before, there is considerable interest in military applications, due partly to the general revival of funding in this sector and partly to the needs of new weapons systems. We can expect this interest to continue for some time, and to lead to research and development in topographic applications of GIS technology, and particularly in methods of editing, verifying and updating dense Digital Elevation Models (DEM). Some of the proposed work in Artificial Intelligence (AI) for Strategic Defence Initiative (SDI) will have relevance to GIS because of its concern for image processing, and there may be other connections between GIS and strategic weapons systems.

Previous discussion identified marketing and related applications as a major growth area. Research and development in this application field is likely to be funded by the private sector. However the field has intersections with a number of government and municipal activities, particularly statistics, public works, crown corporations in transportation, education and emergency facility management, and small amounts of funding may find their way into GIS research and development through such channels. Also there is an increasing tendency for the national basic research funding agencies, National Science Foundations, Natural Sciences Engineering and Research Council of Canada, Social Sciences and Humanities Research Council of Canada, etc. to favour programs which match money raised in the private sector.

3.4   NATURE OF GIS RESEARCH

Over the past 20 years the roles of GIS, automated cartography and computer aided design (CAD) have frequently been confused, both in the relative applicability of each technology in various application areas, and in the direction of basic research. The respective data structures have very little in common; the GIS literature makes very little reference to automatic cartography or to computer aided design; and whole areas which are intimately related to GIS, such as spatial analysis and spatial statistics, have no relevance to automatic cartography or to CAD at all. On the other hand the development of digitizer and display technology has clearly benefitted from the influence of the much larger CAD market.

We believe that GIS is a unique field with its own set of research problems, and although we would be willing to defend that position, and argue that it is the logical conclusion of 20 years of North American experience, we would not expect the entire GIS community to agree with it at this time. A GIS is a tool for manipulation and analysis of spatial data; it is therefore in the same relationship to spatial analysis that standard statistical packages such as SAS and SPSS are to statistical analysis. This is radically different from the purpose of CAD or automated cartography, and has led to the development of fundamentally different data structures and approaches.

If follows that the set of potential applications of GIS is enormous, and is not satisfied by any other type of software. Future GIS development will depend on better algorithms and data structures, and continuing improvements in hardware. But is also needs research effort in spatial analysis, in the development of better methods of manipulating and analysing spatial data, and in our understanding of the nature of spatial data itself through such issues as generalization, accuracy and error. Thus GIS research needs to be concentrated in three areas; first, data structures and algorithms, second, spatial analysis and third, spatial statistics. Hardware development will likely continue to be motivated by larger markets in computer graphics and CAD.

4 Institutional concerns

Several of the issues which arise in this context have already been addressed, but will be reiterated here in the form of 'lessons we have learned'.

First, North American political structures make it difficult to organize coordinated approaches to the development of technologies like GIS. In Canada the greatest success has been at the provincial level, but even there there are consistent patterns of duplication and lack of planning. The North American suspicion of big government and bureaucracy also operates against attempts to establish large, well-coordinated spatial data bases. The mapping function is split over many agencies and over several levels of government. The result has been a haphazard approach in which such standards as have emerged have come about because of the small number of vendors in the market, rather than through any central coordination.

Second, there is an increasing gap between the personnel needs of agencies which have acquired GIS's and the ability of the educational system to provide them. Again this stems partly from the lack of central planning in the educational system and the level of autonomy enjoyed by most educational institutions. A process of natural selection is currently occurring, among both institutions and disciplines, and it is too early to tell what the result will be.

Third, GIS is a highly attractive technology, and suffers from the problems of all such technologies. It has been repeatedly wrongly sold as a solution in response to needs which were poorly defined or not defined at all, and to clients who did not really understand its capabilities or limitations. Many failures have resulted from the acquisition of the wrong type of system, because of poor advice or lack of advice. The success of the technology is in many ways dependent of the availability of good advice, through either public sector centres of excellence or competent private sector consulting.

Fourth, many failures in North America have resulted from agencies attempting to develop their own systems with inadequate resources and exaggerated goals. The systems currently on the market only partially satisfy the notion of a GIS and yet typically contain 10 or more instructions, and represent investments of several man-years of programmer time. Yet there have been many examples of agencies embarking on projects to develop similar functionality from scratch. Such inhouse public sector R&D would be of some benefit to society at large if it resulted in substantial cooperation with the private sector and eventual sales, but the record on this in North America has not been good: instead, public sector development have tended to compete with the private sector, with all of the attendant problems of hidden subsidies and unsatisfactory tendering procedures.

Finally, there are just as many problems, and possibly more, on the management side of implementing an information system as there are on the technical side. As we have already seen, a primary benefit of a GIS lies in the new capabilities which it introduces, rather than in the ways in which it allows old tasks to

be done more efficiently or more cheaply. To be successful, then, it requires strong and consistent motivation on the part of all users, the great majority of whom will have no technical understanding of the system. We have learnt that such motivation will not occur naturally, however user friendly the system, however impressive its products, and however sophisticated the functionality. It is absolutely essential that the users of the system be the ones who plan it and arrange for it to be acquired. Systems which have been installed on a trial or experimental basis, 'to allow potential users to see the benefits', almost invariably fail once the honeymoon period is over.

A successful GIS program at the national scale would seem to need the coordination of five types of effort. First, it would need a set of valid application areas. Based on our experience in North America there seems at the moment no reason to doubt that such areas exist and that they will grow rapidly in the near future. Second, it requires a set of active vendors. Again there is no reason to doubt that the set of available products will grow or become more sophisticated in the next few years. The size of any nation's share of the international pie is likely to depend partly on the size of the nation's application areas, but primarily on the remaining three types of effort.

The third is the educational sector, which will provide the trained personnel to run systems, conduct basic research, staff the vendors and train future generations. Although various North American governments have funded programs in various universities and institutes, the total 'top-down' influence remains small and most courses and programs exist because of a 'bottom-up' perception of a demand for graduates.

The fourth type of effort is in research and development. It seems from past experience that most major research breakthroughs will occur in governments and universities rather than in the private sector; commercial incentives appear to have led to successful adapting and improving, but not to major new directions, and the most significant advances in the GIS field remain those originally made in CGIS and subsequently at the Harvard Laboratory for Computer Graphics and Spatial Analysis. This means that government will remain the major source of funding, and the major control on the rate at which the field progresses.

Finally, because of the managerial issues which we have already discussed, successful application of GIS requires not only the four types of effort already mentioned, but also the existence of a substantial source of expertise which is independent of vendors and acquiring agencies, and capable of mediating the acquisition process and ensuring that it leads ultimately to success. This role can be filled by an independent, commercial consulting sector, but it is unlikely that this could exist in the early, critical period of national GIS involvement. It is a role which could be filled by a government centre of expertise, which would have to rely to some extent on imported talent. It is something which we have found to be critically lacking in developing countries, and in industrial countries which have come into the GIS field rather late because of the small size of the local market.

Appendix 7
Spatial units and locational referencing

The Committee of Enquiry invited Dr S Openshaw MBCS, FIS of the Centre for Urban and Regional Development Studies, University of Newcastle Upon Tyne, to provide a review study on Spatial Units and Locational Referencing. The views expressed in this paper are the personal views of Dr Openshaw. The paper was drafted in 1985.

Introduction

1   This is a review paper with the following terms of reference: a commentary on previous recommendations and current proposals for locational referencing, a review of the alternatives, some indication of possible developments in the next 5 to 10 years, and an outline of issues to look for in the analysis of submitted evidence. Section 1 covers some of the more general issues. Sections 2 and 3 provide a critical review of various spatial units and locational referencing schemes. Possible future developments are discussed in Section 4.

2   The processing and availability of geographically referenced data are rapidly emerging as very important areas for both government and private sector organisations. Increasing amounts of data are becoming available on computer media, the value and usefulness of which may be greatly improved by geographic handling techniques. These data are often produced as a result of routine management processes; however, their potential value to third parties will largely depend on the locational referencing systems they use, or do not use. Examples include: in central government, census data, tax returns, Department of Health and Social Security (DHSS) files, and crime data; in local government various administrative files particularly in areas related to rates, planning, electoral registers, social services, and education; the public utilities and regional health authorities possess much useful data; and the private sector collects a wide range of personal information.

3   Currently, many of these data sets are confidential, they are now protected by the 1984 Data Protection Act, and historically have generally not been shared between users. However, it is apparent that these data sources have considerable value both for administrative and commercial purposes to third parties if they were to be available and cross-referenced on an areal basis. To meet these more general needs, the data would have to be locationally referenced so that a high degree of aggregational flexibility was possible. At present the potential opportunities in this area are severely limited by the use of fairly unique and non-flexible sets of areal units and a correspondingly low degree, or absence, of aggregational flexibility.

4   Most of the problems are not new but their overall importance has grown tremendously in recent years, in line with the vast data explosion that has accompanied the nearly complete computerisation of the principal services and administrative functions necessary for modern life. Vast amounts of hitherto inaccessible data are now available in computer readable form. At the same time the costs of processing and manipulating these data have been dramatically reduced by advances in computer hardware. Many technical obstacles that hitherto precluded secondary analysis of data collected for routine administrative purposes have now gone. Instead the wider usefulness of many files is now dependent on their availability for analysis (an institutional matter) and on the adoption of efficient, standardised forms of locational referencing. The latter is necessary to allow for the creation of multi-source geographically referenced information systems that permit data to be assembled and reported for virtually any set of spatial units. This is an important prerequisite if there is ever to be a national information system covering all aspects of modern life. The Committee may well wish to consider whether such an outcome is implicit in the recommendations they may make, and whether this long-term objective needs to be explicitly identified.

5   Geographic information handling is a crucial part of this process. A geographic area base is necessary for matching different data files because of the absence of anything resembling a person specific, unique, fixed, computer key or identification number. It also offers a way of cross-referencing personal data without breaching confidentiality restrictions; since it is the geographic areas not persons that are being cross-referenced. The Committee should be aware that the potential value of such cross-referenced data may often far exceed the sum of the value of the original data sets used to create it. This applies to data used for administrative purposes as well as in the private sector. However, the usefulness of the data will be dependent on the size of the spatial units used and the number of different geographic areas that the data can be reported by. Government may be quite happy with fairly coarse administrative areas but many local government and private sector agencies will often require data to be assembled for far smaller areas. It is quite apparent that without a high degree of aggregational flexibility, third party usage will remain low. Modern geographic information handling techniques offer the prospect of a high degree of aggregational flexibility.

6   It is important, therefore, that the most careful thought be given to how the current opportunity for specifying locational referencing standards and in explaining to others the potential value that can be added to their data, should be handled. A similar opportunity existed in the early 1970's just before the beginning of various large scale computerisation exercises of government administration at both local and national levels. On that occasion, the recommendations were ineffective and as a result several important streams of information remain unavailable for analysis. Decision making in local government is not as well informed as it might otherwise have been and the development of central government databases for corporate use have been hindered. Additionally, important data series have developed discontinuities because of the instability of the spatial

units used. Finally, very little progress has been made towards both encouraging the secondary analysis of many important data sources by third parties and of exploiting various markets for government data sources at a sub-standard level. It also seems that public money has been spent in trying to handle by clerical operations data manipulations that should have been performed by geographic information handling in an entirely automatic fashion. The risks for the future are considerably greater and without appropriate action the full benefits of the new opportunities created by the information revolution may be unavailable, to everyone's disbenefit.

Section 1: Issues

7   It is useful to precede a more detailed review by identifying some of the important issues to be kept in mind when considering alternative spatial units and locational referencing systems. It is assumed that the prime reason for wishing to use geographic information handling techniques is to try and enhance the wider usefulness of data, although the degree of added value will be very variable and depend on such factors as the nature of the locational referencing that is used, the demand for the data, and the opportunities that exist for cross-referencing different data sets by the same sets of areas.

8   The following major issues can be identified:

1   What are the objectives in encouraging the wider use of geographic information processing and why should these be relevant to government?

2   What sorts of spatial units are in common usage, where did they originate, who uses them, how fixed are they, are some more common than others, and what data are available for them?

3   Is there a common perception amongst data providers and users of any problems that result from the use of multiple different and incompatible sets of spatial units, and how can these problems be solved?

4   Is there a demand and market for multi-user databases for areas derived from many different sources, and if there is, how sensitive is it to the nature of the spatial units used to report the data?

5   Do de facto locational referencing standards exist, if they do are they the most appropriate ones for the future?

6   To what extent can appropriate locational referencing add value to data held by central and local government and is there a market demand for such data?

7   What benefits might accrue to local and central government and are their perceptions of gain sufficiently strong to justify the overheads involved?

8   Will the 1984 Data Protection Act's restrictions on the handling of personal data increase the importance of cross-referencing data by areas?

9   Is there a demand for time series data for fixed areas and what is needed to meet this demand?; and

10 Who will be the main users of geographic information processing and to what extent can these activities be centrally organised?

9   In addition, a number of more general issues can be identified. What would happen if current trends continued, which areas of concern might be most sensitive to the Committee's influence, and what realistically can the Committee expect to achieve? To what extent should the Committee be focusing on the specifications of locational referencing standards, educating data providers and data users, providing an infrastructure to allow the more effective use of geographic information handling within government (both local and central), and encouraging the creation of national databases and Geographic Information Systems as a service to both government and the private sector?

10 A final comment, it would be naive and unrealistic to expect that meaningful monetary values can be attached to the likely 'costs' and 'benefits' associated with geographic information handling. A large measure of 'faith' is needed to start the process off.

Section 2: Spatial units as a simple form of fixed locational referencing

11 A Basic Spatial Unit (BSU) is generally used to refer to the smallest spatial entity to which data are coded. Ideally, this should be small enough and common enough to be a universal constant across all data files. In practice, there is no such definition and the nature of the BSU depends on the entities to which the data refer; for example, a house, a grid-square, an administrative area, and a postcode. To complicate things further, three different sorts of BSU can be recognised; (1) that spatial entity for which data are collected; (2) that for which it is stored in a computer file; and (3) that which is used to report the results. All three could be the same but usually the BSUs used to report data are not the same as those used for data collection and storage. Herein lie many problems. For example, data about people coded only by standard region have very little aggregational flexibility. The same data stored coded by unitary postcode can be aggregated to standard regions and to many other sets of areas as well. The nature and geographic resolution of the locational referencing attached to data files is, therefore, a very critical factor in the subsequent re-use of the data, as is also the size and nature of the spatial entities which are used as the basic units for data storage.

12 Geographic locations can be represented as zone codes. At best they convey an impression of approximate location (viz. within a particular postal area). If the zone codes are hierarchically structured then certain simple sorts of geographic information handling can be performed automatically; for example, you can aggregate census enumeration districts to wards, wards to districts, and districts to counties. However, the range of possible aggregations is fixed by the zone codes that are used; you cannot aggregate postcodes to census enumeration districts without additional information not contained in the zone codes.

13 The maximum degree of aggregational flexibility requires the presence of a national grid-reference (or some other consistently defined coordinate system) but even then whether an aggregation can be performed precisely or approximately depends on the extent to which one (or more) grid-references can accurately represent the BSUs being aggregated. If the BSU is small in size then it can be usefully represented by a single point reference, provided the spatial resolution of the reference is sufficiently precise. If the BSU is small but the resolution of the grid-reference (ie number of digits) is poor (eg a 1km reference for a house in a town) then there will be problems when aggregating the data. If the BSU is large then it may be necessary

to record not one but several grid-references to represent its boundary, unless the BSU is completely regular (ie a grid-square) or a linear feature (ie a road) in which case a different approach is required.

14   Spatial entities which have one or more coordinate references to code their geographic location are far easier to process, including the automatic aggregation of data from one set of BSUs to another, larger, set of areas. This form of locational referencing greatly increases aggregational flexibility since the range of possible aggregations are not pre-defined by the zone coding scheme used. Processing costs are higher but not significantly so; for example, it would not be particularly expensive to assign all 1.6 million postcodes in Britain to county boundaries, if it were done in a sensible fashion using state of the art techniques. However, not all aggregations will be completely accurate. Unless the BSUs being aggregated and the areas to which they are being aggregated nest into each other, there will be incorrect assignments along the boundaries (ie edge effects). For example, postcoded data can be assigned to local government districts but with some slight errors due to the fundamental incompatibility of postcode and district boundaries. If boundaries were available for both, then the problem postcodes could be identified albeit at some greater computational expense.

15   Mismatches between the boundaries of spatial units of different types are common and can seldom be eradicated. In practice a degree of error has to be accepted as a fact of life and either ignored or some kind of 'best fitting' scheme adopted. The greater the size difference between the spatial entities being aggregated and the areas to which they are being aggregated, the less the errors matter. There may well be an indifference curve, given the error-prone nature of all data.

16   Finally, it is noted that there is no such thing as a universal BSU. Different BSUs are relevant to different data sets and different users. The only practical way of cross-referencing spatial data in general is by the use of the smallest practical BSUs. Ideally, these should be representative of the entities about which data are recorded; for example, for personal data the postcode, for remote sensing data a grid-square representation of sensor pixels.

SPATIAL UNITS AS LOCATION CODES
17   The most common form of simple geographic referencing system is the use of a fixed set of areas for both coding and reporting spatial data, area codes are described via a code-book just like any other variable. There are many fixed spatial reporting frameworks in current use. Some of these spatial units can be related in various hierarchical ways because the BSUs nest within each other; for example, data for Districts can be aggregated to Counties and Standard Regions. However, usually the range of possible data aggregations that can be peformed are very limited and they are difficult to automate without first devising, by manual means, indexes that relate one set of spatial units to others. It is often very costly, or impossible, to change data that have been defined for one set of BSUs to a different areal base. For example, the 1971 Census of Population used a largely different (55 per cent or so) set of enumeration district definitions than those used for the 1981 census; this precluded any detailed analysis of change at this finest level of resolution and necessitated the hasty construction of a set of census tracts as an intermediate scale to which both 1971 and 1981 data could be aggregated.

18   The magnitude of the data incompatibility problems resulting from the use of a large number of different and geographically incompatible sets of spatial units is still increasing due to the total absence of any standards and as more data become available. To some extent these are all potential rather than real problems, because they only occur if you try to bring together multisource data sets. It is possible that major public sector savings could have been achieved had there been an early standardisation of the spatial units used for storing and reporting statistics for central and local government. This does not imply the choice of only one set of areas, merely a set of rules and standards for locational coding that are common across Government Departments. The problems mainly manifest themselves in the non-availability of data that are known to exist but cannot be used because of the nature of the spatial units used for coding location. In practice, this will seldom be identified as a problem because the system is so good at hiding its extent: thus data are not considered important because they cannot be accessed for an appropriate set of spatial units. The true extent of these spatial area incompatibility problems tends to be hidden.

19   Whilst central government may be fairly satisfied with the range of data obtained for local government administrative areas, they are seemingly not concerned that the data provision reflects the compartmentalisation of Government Departments; for example, the Department of Environment (DOE) gathers statistics by local authority areas, the Department of Employment (DEmp) by travel to work areas, the DHSS by health regions and health districts, the Home Office by Police Divisions and Force Areas etc, none of which are either comparable or for the most part nestable. This is not perceived as a problem because the data serve the needs of the operational units involved and data provision to secondary users is of less importance. It seems that any notion of a corporate, common, database for government at either national or local levels is still an alien concept; it should not be, but it is. The Committee may like to consider the implications of this state of data disorganisation for government because it is here that one of the principal uses for geographic information handling exists.

20   Perhaps as a consequence of these problems, it is hardly surprising that there appears to be little use made by Government Departments of each others data. This may be a little severe because of the absence of any published information about such usages. Nevertheless, some of the data sources collected by one Department would seem to be of considerable interest to others; for example, tax returns could be used to provide data about incomes and this might be very useful for measuring the needs element in the rate support grant, if such data could be reported by local authorities. Obviously the individual tax records would be highly confidential information but when reported by districts or counties or even wards and census enumeration districts confidentiality could be assured by applying current Central Statistical Office (CSO) restrictions. There might even be a large private sector demand for the data that would more than cover the data processing costs. The various Rayner reports seemed to have missed some of these aspects.

21   There are many sets of different spatial units in common usage. Some appear to have no apparent relationship or relevance to anything and seem to survive only because of the problems of changing them. This criticism is relevant only to the extent that the very nature of the spatial units can influence the data they generate and affect decision-making based on the

results; this is discussed later as the modifiable areal unit problem. There is, perhaps, a more general educational point here that the Committee may like to consider. It is quite likely that there may well be various organisational advantages to using more meaningful spatial units for planning purposes. It is clearly not the Committee's responsibility to suggest wholesale changes in organisational structures which are reflected in the spatial units that are used, but they may wish to draw attention to these aspects.

22   Other problems occur when data relating to small areas are coded only for those coarse areas used for reporting purposes. For example, every police force has a well developed computer information system designed to record crimes according to Home Office specifications. However, the Home Office is only interested in aggregate statistics for entire Forces and Divisions, so even though data are stored for each recorded crime, the locational codes are often to Force and Division. As a result a vast amount of potentially very useful data are collected but cannot be used at a geographic scale where analysis would be most helpful; ie by subdivision and more especially police beat. Most police forces simply do not know anything about the detailed pattern of crime for small areas even though they already collect all the necessary information. Suppose, however, that a Force improves its locational coding so as to reference police beats (the smallest area of any interest to them). Analysis probably requires matching crime data for police beats with census data so that crime rates can be computed. Unfortunately, police beats seldom, if ever, match census enumeration districts; nor are they easily expressed in terms of them, nor is it possible to re-assemble the crime data by census enumeration districts. Linkages with other potentially relevant data sets (held by local authorities) are likewise either impossible or exceptionally difficult to manage and at best rather approximate.

23   Again, does it matter that data are not usable for purposes which are somewhat different from that for which the information was originally required? What is the cost in terms of possibly inefficient planning and management decisions? Are techniques available to help management cope with a far more detailed supply of data, and would managers welcome them? For example, does it matter that British Gas, the CEGB, and the NCB all have different planning regions and thus are unable to assess their market shares at anything other than the national level? Likewise, subnational energy forecasts are only possible if each energy utility makes its own independent forecasts of the performance of their market competitors. Does it matter that national government do not have ready access to all the relevant data for even the administrative areas in which they have an interest? Finally, to what extent do private and public sector organisations have an accurate appreciation of the total supply of data, including that held outside their immediate area of operations, that may be of interest to them? In the information era does there need to be a 'freedom of public data' Act that will encourage the sharing and pooling of spatial data of common and vital importance? What forms of locational referencing are a prerequisite for this to happen?

MODIFIABLE AREAL UNIT PROBLEMS
24   There is another problem which is common to all spatial units and all spatial data which is usually referred to as the modifiable areal unit problem (MAUP). It concerns the conditional and modifiable nature of the data that are produced when data (originally defined for one set of spatial entities) are aggregated to another. The resulting values depend to some

unknown degree on the spatial units used for the aggregation and are known to be influenced by the nature and configurational properties of these units. For example, the unemployment rate for an area may vary depending on whether the definition is based on administrative boundaries or on a functional region definition or on boundaries carefully designed to maximise (or minimise) the resulting rate. Computer techniques now exist which can design new spatial units (by adding together one or more old areas) so as to explicitly optimise some function based on the data (for example, unemployment rates). These techniques have been used to design new sets of constituency boundaries which are more equal in population size than those used today. It is also a problem in that the use of any set of spatial units inevitably imparts some degree of bias to the data.

25   Techniques for engineering spatial units to match particular purposes are likely to become important and very useful in the future. They would allow new sets of spatial units to be designed which possessed particular properties and which were built from an existing set of areas. This will become necessary when data users realise the extent of the endemic MAUP. It may also offer organisational advantages. For example, the design of planning regions should, ideally, be done in a manner which is directly related to the activities being planned. The CEGB, for instance, make decisions concerning the location of additional capacity using planning regions that are historical accidents and are seemingly unrelated to many relevant variables (ie transmission network, demand centres, and existing supply points). It is possible that, as a result of using inappropriate spatial units, they locate new capacity in less than optimal locations.

26   This line of reasoning can be taken further and used to suggest that a number of standard sets of spatial units, perhaps built from a number of different base areas, should be identified each with known properties, to be used as spatial data reporting frameworks. An example would be the Department of the Environment's use of functional regions in the 1970's; another would be the continuing use of another set of functional areas by the DEmp, Manpower Services Commission, and Department of Industry (DTI) ie the Travel to Work Areas, the current versions of which were defined in 1984. These functional regions are designed to represent the spatial organisation of a heavily urbanised country by identifying labour market areas, each of which possess a minimum size and a minimum degree of commuting self-containment.

27   The Committee should be aware that these MAUP problems exist. The implications for their recommendations concern the need to educate users of spatial units and to ensure that the additional data needed to exploit zone design techniques are available (ie boundary files from which contiguity lists can be generated). They may also wish to consider whether or not the provision of a standard set of designed spatial reporting frameworks appropriate for national level data might be useful, and whether or not 'official' statistical areas should be re-formulated in this way.

ECOLOGICAL FALLACIES
28   Another important feature of all spatial units concerns the dangers of users committing an ecological fallacy whenever data for such units are analysed. It is necessary to remember that the features reported for areas may not provide a good description of the individuals who live there. The spatial aggregation of individual BSU data may itself be creating spurious associations

and, depending on the scale of the spatial units, can result in area profiles which are seemingly distinctive but have in fact been generated by the aggregation of data about quite separate groups of people or households. For example, area type K37 in the ACORN classification of census enumeration districts is described as 'private houses, well off elderly'; and type K38 as 'private flats with single pensioners'. Yet 65 per cent of K37 and 73 per cent of K38 are below 65 years old. These areas are spatially distinctive because of the relatively high concentration of old people found there, but these are minority characteristics. Thus data for areas may sometimes provide a poor representation of the characteristics of the individuals who live there. For many purposes this does not matter and is even an advantage as a means of preserving individual data confidentiality. It is only a problem if the data user is unaware of this limitation of spatial data.

29   The Committee ought to be aware of the potential harm that a total neglect of the ecological fallacy problem may conceivably inflict on innocent people. The only practical solution is at best partial and involves the use of the smallest possible spatial units. The only other provision that can be made is to urge that individual data be used to establish notional limits on the range of observed effects for each data set. Previously, as for example with the census, it has been totally neglected.

Section 3: Other forms of locational referencing

30   The principal advantages of using fixed spatial units to code locations are those of convenience and simplicity; the geographic referencing can be done automatically (ie. from a postal address) or by clerical action. The disadvantages are those of inflexibility if an area code is the only form of locational reference available. There is need for a more flexible locational referencing system in which the locational reference itself can be used in automatic geographic information handling processes. This is important also because it is unlikely that there will be any single fixed set of BSU's suitable for all users and therefore aggregational flexibility is extremely desirable.

31   It is worth noting that flexible aggregation can greatly enhance the uses to which spatial information can be put and allow hitherto unconnected information to be extracted using geography as a relational mechanism; for example, to identify the data characteristics within the neighbourhood of a specified town. However, complete aggregational freedom is only likely to be obtained if the BSU's are sufficiently small (ie. individual households or very small areas) and if the expense can be afforded or spread over many data users (ie. as with the census). Probably the best way to identify issues and lessons from experience is to briefly examine various proposals and operational Geographic Information Systems related to government.

THE GISP REPORT 1972
32   A DOE report 'General Information Systems for Planning' (GISP) represents the only previous attempt to try and tackle the general problem of how to organise information, including the question of geographic referencing. It recommended the creation of what was referred to as a 'complete' system which amounted to a fairly comprehensive Management Information System (MIS) that would require a corporate approach to data collection, recording, and supply within a Local Authority to help it meet new information needs created by the 1968 Town and Country Planning Act which put considerable emphasis on monitoring activities. GISP originated as a means of organising

planning related data but it was quickly realised that the information needs of the planning department could not be separated from those of the remainder of local government. The time was opportune because of the impending re-organisation of local government and the prevailing enthusiasm for the new form of structure planning. Comparable methods were supposed to be adopted also by central government. It should be emphasised that few of the GISP recommendations were implemented and then only on a small scale. It was an ambitious, pioneering study that failed to have any lasting impact or any significant effect.

33   GISP recommended the use of a common basic spatial unit of comparison which in the built-up area was to be the rating hereditament or some subdivision, elsewhere the Ordnance Survey parcel (as shown on large scale maps), or in less developed areas without large scale maps coverage, the individual field. In addition, there was to be a second tier of spatial data unit, the planning zone. This was to be used when BSU level data could not be made available for confidentiality or other reasons. The definition of planning zones was left to the user and would reflect local circumstances and interests; in practice it would probably have been zones used for computer modelling purposes. It was never clear how data that existed at neither the BSU level (ie rating hereditaments) nor for areas that matched the planning zones was to be handled; presumably it was assumed that all data would exist at the BSU level. The nature of the BSU reflected purely practical considerations; particularly, the availability of address lists for rating hereditaments in urban areas.

34   The spatial units recommended by GISP were meant to completely cover a local authority area so that whatever objects or data may be of interest to any user they would already have a BSU for them. In practice this might well result in very large numbers of pseudo-BSUs being created to cover such entities as traffic islands, pelican crossings, and various other ad hoc items just in case someone wanted to refer to them. Other complications would be caused by the land development process which would progressively change the pattern of local BSUs as the development proceeded.

35   Once spatial coverage was complete, the BSUs would be accessed by giving each of them a unique property reference number (UPRN). An index was to be generated to provide for each BSU the following information: a postal address, a postcode, a UPRN, and a 1 metre national grid-reference to identify a centroid location (ie. a point in the approximate centre of the rating hereditament). The grid-reference was to provide aggregational flexibility by allowing various types of spatial data processing to be performed. This index or gazetteer is very important because it provides the way into the system. A file of data to be added into the system would consist of a series of postal addresses. A clerical operation would be used to look-up each address in the gazetteer so that the appropriate UPRN could be attached. This UPRN would then allow the data to be cross-referenced at the BSU and planning zone levels with any other data that might exist. Additionally, the UPRN link also allows access to a grid-reference which would make possible the aggregation of the data to other sets of areas for which boundary descriptions were available. It was clearly labour intensive in that the task of looking up hundreds of thousands of addresses in a paper gazetteer should not be underestimated. Also for the system to work, the gazetteer had to be prepared: first, it had to be accurate, and it had to be comprehensive so that every conceivable BSU was included and point referenced to avoid

duplication of effort. It was intended that the entire edifice would be updated on a weekly basis.

36 The GISP report does mention area and network referencing but it is not so obvious as to how it could be fitted into the recommended framework.

37 The principle problems with the GISP recommendations are as follows: the complexity of the gazetteer creation and maintenance task was grossly underestimated; the spatial data processing techniques on which it depended were not available until much later; the proposals rested heavily on pioneering and experimental point referencing systems that were either never to become operational or were really manual rather than computer systems (eg. the Coventry point referencing system); the report could only recommend and could not be implemented if local authorities chose not to; it was only a paper specification of one possible system design; it was very ambitious given the available computer hardware in local government; and it was only really appropriate to metropolitan and urban areas.

38 Some of the statistics involved in operationalising GISP testify to the enormity of the task. The potential number of URPNs was of the order of 20 million rating hereditaments and several million OS parcels and fields. The task of creating and maintaining this index would have been the responsibility of 458 different local authorities, some without adequate computing facilities. Outside the major urban areas, there was not even an existing register of potential BSUs. GISP required a level of coordination of effort that was quite impractical. It needed a long-term commitment based largely on faith in the need for an information infrastructure based on property, one that would offer very few short-term benefits to those involved.

39 For various reasons, GISP seemed to lose momentum; possibly because the problems for which it offered solutions were not regarded as being sufficiently important to justify the effort involved, and possibly because of delays in funding the 31 follow-up studies. There was also only limited acceptance within local government and it was significant that no computer manufacturer or software house was either asked or was prepared to develop an operational GISP system for the local authority market. The few attempts that were made, were either research orientated (for example, IBM's UMS system) or quite unrelated to GISP (the ICL LAMIS system). The practical benefits were unclear and there was no statutory requirement to implement either GISP or to conform to its standards. On the other hand, there is little doubt that GISP did generate considerable interest in computerised planning information systems, but it failed to provide a practical system.

The national gazetteer pilot study, latterly the joint information system of Tyne and Wear

40 For GISP to work efficiently at all levels of local and central government it was recognised that there had to be a single common national gazetteer, in order to avoid unnecessary duplication of effort and to establish a common BSU framework for the entire country to be applied to all data. Accordingly, in 1973 the National Gazetteer Pilot Study (NGPS) was funded to investigate the problems of gazetteer creation in the non-ideal multi-local authority of Tyne and Wear County, so that hopefully a national gazetteer could be established in time for the 1981 census. In 1975, the system was renamed the Tyne and Wear Area Joint Information System (JIS) supported by the five Districts. The first reliable gazetteer was not available until April 1978. This JIS system continues in existence today.

41 As information systems go, this GISP relict is very rudimentary and does not really match the description 'information system'. It is spread over five separate district computers with additional processing being done on the county computer although the County Council is not allowed access to the data held by districts for political reasons. The gazetteer consists of 500,000 BSUs each with a UPRN, a land-use code, a 1 metre grid-reference, a postal address, and a few area indicators. The JIS system is still primarily concerned with maintaining the gazetteer using a monthly updating cycle. It survives because the economic benefits of having an up-to-date property register have been sufficient to convince local politicians to continue and slowly expand the scope of the system. However, JIS is not a single integrated entity with a whole host of advanced (or indeed many) geographic information handling processes. Instead it exists primarily (it differs a little from the more advanced to the least enthusiastic local authorities) as a database still predominantly accessed by batch programs. Yet the very fact of its survival for such a long period in such a hostile and difficult political environment is a tremendous tribute to the diplomatic skills and persuasive powers of the JIS team.

42 If the JIS was to suddenly disappear, it is doubtful whether this would affect the delivery of local authority services. Nevertheless, the Districts are keen for the JIS to continue after the dissolution of Tyne and Wear County and there is every prospect of the system being developed into a full computer information system. At present, its potential for management purposes has yet to be realised because of hardware and software problems. In any case it is suggested that much of the technical experience gained from the NGPS and latterly JIS is not particularly useful or relevant. No one would now try and create from scratch a point referenced gazetteer based system by digitising from dyelines of Ordnance Survey (OS) maps. Commonsense dictates that if such a database is needed then attempts are made to extract all relevant data from OS digital map data structures. Additionally, the expense of creating and maintaining a comprehensive GISP-like gazetteer is hardly worthwhile. The amount of effort is directly proportional to the number of BSUs. GISP was far too ambitious and also unnecessarily complex; most of the potential benefits are now achievable from much simpler postcode based systems. However, to be fair, back in 1972 there was probably no real alternative to the GISP approach given the objectives that were set.

43 The operational experience generated by over a decade of GISP-like functioning by the JIS team is likely to be their most valuable contribution to this Enquiry. The local government view would probably be that the benefits of the GISP style of information system are very limited, although in the case of the JIS system it may well reflect the lack of an integrated computer structure as well as a failure to include all relevant data in the system. JIS certainly indicates that GISP can be made to work and that the real advantages are only likely to become significant when local authority wide corporate databases can be established and run for a number of years. I know of no examples where such systems exist. It is clearly futile to try and build a national Geographic Information System from the bottom up when the local authorities are responsible for their areas. Local authorities are simply the most inappropriate agency for such work and extreme doubts are expressed as to whether there would be sufficient tangible benefits to local government to sustain the development of such a system.

44 It is hard to imagine the magnitude of the benefits required to induce all local authorities at a time of recession and budgetary pressures to apply common locational referencing standards to all their data files. Some would certainly do this and may be well on the way already; the majority would probably refuse. To have any hope of being successful such developments would have to be clearly related to immediate day-to-day management functions or specified as a statutory requirement. Even the GISP report is cautious. It states: ' . . . whereas costs are generally calculable, benefits are only calculable in the minority of cases. For instance, it is scarcely possible to calculate the financial benefits of being able to make better and more timely decisions.' (page xiii, xiv).

45 If the Committee regard local authorities as the principal agents or beneficiaries of improvements in geographic data handling consequent on implementation of their recommendations, then they may well be disappointed by the lack of enthusiasm from that source. Now there are some people who believe, based on intuition and experience, that the benefits are real and realisable but they need to be demonstrated in an unequivocable fashion to hard-nosed local government managers. The problem is not purely that of data provision but it also involves data use and how the improved and enriched supply of data are to be used to make 'better' decisions.

LAMIS

46 The Local Authority Management Information System (LAMIS) was developed at the same time as the GISP proposals but seemingly in ignorance of the GISP recommendations. LAMIS was the result of a joint project by Leeds, ICL, and the DTI; it was viewed by government as a means of supporting the sales of ICL computers. LAMIS differs from GISP in that it was conceived as an operational computer system (ie. a software package) that offered a standard set of database facilities. It provides for full boundary referencing of both BSUs and zones. Instead of using a single centroid point reference to represent a BSU the boundary of the BSU was digitised. This greater precision in representing location offered the possibility of cross-referencing BSUs via their boundary descriptions thereby avoiding the GISP reliance on UPRNs. This would also appear to allow for the possibility of a direct linkage with OS digital map data structures. Clearly LAMIS offers a number of useful facilities that would seem to allow for advanced GIS operations; for example, updating housing land and development control files via a polygon overlay operation. However, its range of spatial data processing techniques is not sufficiently comprehensive for it to be regarded as a real GIS. Moreover, and far more importantly, the viability of LAMIS in a local authority environment is uncertain. The JIS experience has shown that keeping the system going is far more difficult than setting it up. Its long term utility will be dependent on the accuracy of the boundary files as created and as generated by spatial data processing given that planning data are notoriously fuzzy. Also important are the perceived economics of its operation and its ability to meet user needs.

47 In the world of local government financial assessment procedures it would seem that the economic justification is probably the weakest link. Leeds (the pioneers of LAMIS) apparently decided to abandon boundary referencing on the grounds that the cost of processing such data amounted to over half of total LAMIS processing costs. One imagines also that the progressive build-up of garbage data might well have been another factor. Yet LAMIS without its geographic data handling utilities is no different from scores of other management information systems. The fundamental question concerns what exactly do local authorities require now and in the future? Common sense suggests that their needs should be of a generic nature capable of easy identification. Once found, then universal systems should be possible. In practice this has not happened to any great extent and the reliance on general purpose systems has been substituted for the absence of more specific tailor-made planning orientated systems. It is very relevant, therefore, to question whether or not the planning system needs such elaborate 'all singing and dancing systems' for their geographic data processing. The Committee might be advised to consider commissioning an expert report in this specific area.

48 Finally, it is possible that systems such as LAMIS have tended to foster a concentration on the least useful and most difficult aspects of GIS digital map technology, whilst under-valuing the role of far simpler techniques for handling the geographic referencing of postal address coded data. The issue is again that of defining meaningful objectives.

GRID-SQUARES AND RASTER DATA

49 The use of locationally arbitrary grid-squares attracted considerable interest in the late 1960's as a solution to spatial unit incompatibilities and the need for locational referencing. In the limit when the squares are very small, this is broadly the same as one metre point referencing, except that the objects being represented are squares. For example, a house could be point referenced by a single point but a grid-square representation at an equivalent level of resolution would identify all squares that lay within the boundary of the house building block plan. If the squares were small enough then the resulting raster representation would be equivalent to full boundary referencing, even if it would be a very inefficient way to store the same amount of information. Usually, a considerable degree of spatial resolution is lost in aggregating data to grid-squares; for example, houses referenced by a single one metre point reference might well be aggregated to 1km squares. This illustrates another feature of grid-squares data, namely edge effects. If the spatial objects being aggregated are not themselves squares that are totally compatible with the grid-squares being used (ie. nest perfectly into them), then their assignment to grid-squares cannot be accurate even if the errors will often be trivial.

50 The 1971 census provided statistics for 100 metre and 1km squares. Only the latter were ever of any use due to confidentiality restrictions which supressed most of the finer resolution data. Plans for the provision of similarly constituted 1981 grid-squares data were dropped. However, grid-square data were available from the 1981 census for those areas for which orders were received (about 20 per cent of England and Wales) but the method of estimation was different from that used in 1971, the data were very unattractively expensive due to the charging policy that was adopted, and they were provided for only part of the country (not Scotland) and areas not ordering it. This is a pity because accurate 1km census data would have allowed detailed analysis of 1971-81 changes for a comparable set of identically sized areas, and as an up-to-date population grid for use in computer modelling of the environmental effects of various pollutants. Apart from census data, there are currently only a small number of statistics available on a grid-square basis; for example the PINGAR data on oil product consumption of the Department of Energy and environmental data held by the Institute of Terrestrial Ecology.

51 The principal advantages of grid-squares are their geometric regularity and independence. Provided the squares are not too large, they offer an arbitrary but geometrically regular set of spatial units that will help limit the extent of the modifiable areal unit problem since their shape is fixed. In addition, they offer significant computational advantages when processing large data sets; for example, for spatial searches and for computer modelling in which a grid is essential in numerical analysis. Grid-square data are also easy to map and there are now a number of highly efficient data structures for storage and retrieval. It is also possible to estimate missing data on a grid-square lattice. Finally, they offer a good basis for integrating human and physical data sets. The Office of Population Censuses and Surveys (OPCS) seem to have grossly underestimated the usefulness of grid-squares as a basis for reporting census data and of the potential market for a small number of grid-square count variables.

52 The principal disadvantages concern the inaccuracies in assigning data to grid-squares, their independence of all other sets of spatial units, and the need to use a number of grid-squares of differing sizes. Yet despite these problems, grid-squares are easy to use and it offers a cheap approach to data capture. Map patterns can be rasterised by automatic methods. Furthermore, it is fairly easy to collect data by grid-squares; for example, the DOE's interest in urban land conversion rates could have been handled by asking local authorities to categorise all 1km squares by dominant land-use as an alternative to large scale air photography. Grid-squares also offer a simple approach to developing an area based planning information system. There was some interest in the mid-1970's among Shire Counties in developing a grid-square based information system by converting all their potential digitised boundary data (eg. planning applications, policy regions, map registers) onto a grid. The resulting database would provide strategic information but without having to build and maintain a complex data infrastructure.

53 Finally, it should be noted that grid-squares or raster databases are becoming very fashionable partly because of remote sensing and partly because of their suitability for parallel processing hardware and the application of artificial intelligence. Furthermore, the BBC's Domesday Video Disk, due to be on the market late in 1986, makes extensive use of grid-square technology both for data storage and graphics. It is also intended that this video disk will offer data for about 30 different sets of spatial units covering a very wide range of areas and data sources. It will probably preempt some of the decisions and recommendations likely to be made by this Committee.

NETWORK REFERENCING

54 An alternative basis for describing locations is provided by network referencing. This is mainly useful for representing linear features such as roads, pipes, overhead transmission lines etc. However, it is increasingly being advocated as an alternative framework for the storage of land-use and other property based data that can be related to a network; for example, TRAMS (Transport Referencing and Mapping System) has been used to relate point data to a road network. Other systems offer facilities for surface modelling for road design; for example, MOSS; there are links here with both OS digital mapping and air photometry. The MOSS software seems to have a broad use by County Councils and it is being used to reference planning data. Network referencing can be very effective but it offers a rather inflexible basis for the general locational referencing problem and it is best suited for describing network features which cannot

be easily handled in any other way. If it is necessary to tag other data by network features then this is best done subsequently rather than storing all data in that way.

POSTCODES AND THE OPCS 1991 CENSUS PROPOSALS

55 The GISP report dismissed postcodes as being of very limited value because they were then purely a set of spatial units. Today the situation is very different. Postcodes are widely accepted and used because they have been transformed into a locational referencing system of considerable flexibility. The critical factor was the development in the late 1970's of a postcode-to-100 metre point reference index. This allows locational referencing of all postal addresses for which there is a postcode and it opens up vast amounts of information that were previously inaccessible to geographic information handling. The means now exist to aggregate postcoded address files to any set of spatial units for which there is a digital boundary description. This process could now be applied to databases covering all households and be completed in one over-night computer run (if appropriate data structures were to be used). Postcodes solve the hitherto expensive problem of zone coding address information. It is hardly surprising that OPCS were proposing to create a postcode based geographic coding and mapping system for use with the 1991 census. To a large degree this reflects the commercial importance of postcodes as a means of linking client data with 1981 census data and the utility of postcodes as spatial units for credit scoring and on-line access to personal data. The recent cancellation of the OPCS postcode proposals is a retrograde step, although the OPCS are still examining other ways of offering postcodes as a geographic base for small area statistics from the 1991 census.

56 A single postcode will identify either one or a number of postal addresses (or delivery points) which normally form part of a postman's walk. Small user postcodes (about 1.5 million) consist of an average of 16 delivery points; large users (about 0.18 million) receive over 20 items of mail per day and are assigned a unique postcode; some domestic properties will fall into this large users category. The 1.5 million postcodes can be aggregated to form 8,880 sectors, 2,690 districts, and 120 areas. These spatial units are purpose designed for postal sorting reasons and they bear no relationship with any other sets of areal units. Moreover, bearing in mind their potential wider utility, postcodes provide a very poor representation of all geographic objects which are not postal addresses; for example, parks, open space, building sites; and this may affect their all round utility as general purpose BSUs. GISP avoided these problems by inventing pseudo-rating hereditaments. On the other hand, is it reasonable to expect that any BSU will be able to handle everything? Maybe the solution is to simply assign residual features to whichever address is nearest.

57 Postcodes offer what may be regarded as a simplified GISP-like national gazetteer but with the BSUs being one or more properties rather than an individual property. The Post Office holds a list of all addresses with their postcodes: the postal address file (PAF); and this completes the gazetteer analogy although of course postcodes pre-date GISP.

58 The original OPCS postcode proposals were highly significant in that they would have greatly enhanced the accuracy of the existing postcode-grid reference link by providing digital representations of postcode boundaries. Indeed, because of the massive importance of the census, it

would have established (in reality confirmed) a de facto standard for address based geographic information systems. Additionally, the availability of 1991 census data based on such small areas would presumably have increased the commercial relevance of census data and greatly enhanced the usefulness of all data held by postcodes in various government and private sector organisations. It may also offer an alternative route to the acquisition of census-like data without the need for any more national censuses. Quite simply much of the current census data could probably be obtained from cross-referencing existing files held by government at the postcode level, although this is not part of the OPCS proposals.

59   Finally, it is noted that cross-referencing personal data files held by postcodes is likely to become an increasingly attractive proposition as a means of avoiding (or complying with) the restrictions on the use of personal data imposed by the 1984 Data Protection Act.

60   There are a number of problems with postcodes that the Committee should be aware of. The usefulness of postcodes as a BSU is weakest in rural areas. It is claimed that the maximum distance between farthermost delivery points is limited to 15 minutes travel time; in urban areas it would be a fraction of this. There is also a need to align postcode boundaries to match various administrative areas, or vice versa, and there are doubts about their temporal stability. It should be remembered that postcodes were devised by the Post Office for optimising the sorting and delivery of mail. They reflect the density of postal traffic and as a result the handling of different types of properties are inconsistent; some business establishments will have separate postcodes, others will not. More seriously, the Post Office does not regard postcodes as being geographic areas with fixed boundaries. They are merely a collection of addresses (or one address) to which mail is delivered. Any boundaries that may be recognised are therefore artificial and transient. They could change at almost any time as a result of the three month update cycle that is used. Moreover, large users can retain their postcodes even if they move within the same postcode sector. Clearly, if the intention is that postcodes are to become the national BSU for address related data, then it is essential that some control is exerted over their definition and stability.

61   There may also be a case for replacing the proposed digitisation of postcode boundaries by a digitisation of individual entries in the PAF. This file consists of house number ranges (eg Smith Street, odd, 7-19) and if the start and end houses were to be point referenced then grid-references for the remainder could be interpolated and a boundary description for the complete postcode inferred. In this way the PAF could be converted into a national gazetteer. This level of detail may not matter for many applications in urban areas, although it would be very useful in rural areas and in handling mismatches between postcodes and other spatial units. There is a need to be able to study the accurate geographic distribution of individual houses for which certain information is available. For example, the analysis of children with leukemia in the Black Report relied on postcode based grid-references to generate a ward based data set. A small number of cases are now known to have been incorrectly assigned. It is likely that routine geographic analysis of similar very significant data would require more accurate locational referencing than that afforded by postcodes, particularly in rural areas. Whether this need is sufficient to justify the additional costs has yet to be determined.

62   Finally, the question of handling the updating of the postcode-grid reference index needs careful attention. It could be frozen but this would increasingly reduce its usefulness due to development processes. On the other hand a constantly changing index could only really be tolerated if linkage information were available about the fission of postcode members so that temporal continuity could be maintained. This would probably require that the Post Office accepted that their postcodes had a wider use other than for sorting mail and that they take the appropriate action to support these third party applications. New postcodes would have to be accompanied by either a digital boundary file or a point reference; likewise changes to the membership of existing postcodes would require similar update information to be available.

Section 4: Likely future developments

63   There is likely to be a tremendous increase in the processing of address related information both by various Government Departments and commercial organisations, especially as postcodes are making it easier to do so. It should be noted that various efficient manual, semi-automatic, and fully automated address postcoding systems exist and are being used to complete postcoding of address files that are not currently fully postcoded. The immediate justification is often the need to rationalise address files (ie. to remove duplicate entries etc.) and to obtain discount postal rates when postcodes are used. In theory, large amounts of hitherto largely inaccessible and fairly useless data held for adminstrative purposes are now available, or becoming available, for secondary analysis.

64   For reasons already discussed there is likely to be a growing emphasis on matching files at a fine spatial level, particularly by postcodes, and an increasing trend for commercial users to share and pool their proprietary information (ie. it already occurs for credit rating). This assumes a growing desire and need to make use of data that exists but is currently not analysed by small areas. Some of the potential applications will be research orientated: for example, detailed origin-destination studies of telephone traffic; others will be to exploit commercial markets: for example, details of energy consumption by households might be a good proxy for income; and yet other applications will be to support more general management decisions, in that data utility and data availability are inter-connected and no management can afford not to make use of data they collect.

65   Other data developments concern the growing importance of national databases of geographically referenced information. Currently these are mainly in the private sector but the BBC Domesday Video Disk system gives a hint at what may well become widespread in the near future; namely, multi-source data sets assembled for standard sets of geographic areas with easy access via micro-computer and video technology. It is envisaged that these data systems may well be sold to commercial and governmental users. Their presence reflects the growing realisation by many large national organisations that their hitherto 'useless and nearly impossible to use administrative data sets' do in fact possess considerable market value, once they have been suitably processed. The role of geographic data processing procedures as a means of 'adding value' to data will grow rapidly in the next few years leading to a re-assessment of the utility and value of many hitherto little used data sources.

66   It is suggested that the principal sources of untapped address related data that possess a commercial value are files held by central government and organised centrally. There is really very little data of any value held by local authorities; principal exceptions are probably the register of electors, rating

information were it to be available, and certain planning information. The value of these data are closely related to the provision of similar data on a national basis in a consistent and standardised way; this will be very difficult to achieve.

67 Much of the administrative data held by central government is potentially very useful to a wide range of government functions as well as possessing commercial value. Routine analyses of many currently little or unused data sources from within government will become commonplace once computerisation is complete and personal/property related data postcoded. Subject to the usual CSO type of confidentiality constraints, data from tax returns, VAT, crime, car and land registration, and DHSS will find many useful applications; for example, as census substitutes, as measures of 'need' and 'demand', for informing policy etc. Most of these data are aggregated to and cross-referenced by a wide range of different geographic areas.

68 The growing availability of OS digital maps and the development of software to allow the automatic extraction of data from their data structures could well stimulate the development of very detailed property gazetteers for local areas and open up new developments in GIS. However, it would probably be a mistake to concentrate too much on those developments which involve geographic information being interpreted in the narrowest sense as being primarily cartographic data. The most distinguishing feature about the future is likely to be the range of different requirements and applications. The greatest commercial prospects are probably not in this particular area.

69 Geographic data obtained from remote sensing operations are likely to assume a massive importance creating physical data that was previously either rare or missing, on a routine basis; for example, land cover data, thermal radiation, water content, ground temperature etc. The general utility of these data will be greatly enhanced if it were also possible to integrate them with other socio-economic and land-based data. This would suggest that the development of a suitable, common, grid-square database could be exceptionally useful as a data integrating mechanism with a wide range of users. It is possible that within a decade, many different data users will be wondering how on earth they managed without access to this sort of integrated database. In particular, it may well be feasible to substitute remotely sensed data for data previously obtained in more traditional ways.

70 The importance of grid-squares will certainly increase. Their utility as a time-invariant, stable, set of spatial units which

are common to both physical and human data will be rediscovered. They offer a simple basis for Geographic Information Systems and a number of such systems might be expected to appear. Furthermore, they provide a scale-flexible framework for integrating data from the widest range of possible sources, while their stability is useful for space-time data series with applications in monitoring and computer modelling.

71 The Committee might also find it useful when considering future needs to distinguish between multiple user requirements and those which are predominantly single user or very specialist, albeit important, applications. The level of future market demand may well be stimulated by, and conditional upon, the recommendations that are made. There is a need for the standardisation of data and software, and the market may well be looking for a lead as to which areas are the most interesting for further software development and for pump priming in certain key areas.

72 It is important to recognise that cost effectiveness is crucial. A high cost of access to standard data sets and key indexes will greatly restrict their usefulness and may well encourage the development of simpler, cheaper, and perhaps poorer approaches which in the long term may well be self-defeating. The demand for both geographic data processing and the products created by it may well remain cost sensitive, particularly as many applications will be both new and potentially risky because they are not related to current processes. This is inevitable but, nevertheless, it makes justification difficult. The objective has to be both to identify and establish standards, conventions, and national geographic data resources; and to encourage their adoption and use. One without the other would be disastrous and the key parties are not under the Committee's control.

73 Another major development area is the quickening move to distributed systems. GIS workstations, perhaps using optical disk technology to combine digital and video images, are not far away. The BBC's Domesday interactive video disk offers this kind of facility and it may well establish the feasibility of a prototype national data resource. It is currently intended that data for more than 32 different sets of spatial units, each at ten levels of resolution, will be available for about £300. It will offer a taste of what could be done and will demonstrate the basic principles of what sort of end product is required. The Committee should have a close look at what it offers. The continuing development of really high speed computer networks and file serving machines is likely to revolutionalise the distribution and access to GIS and geographic data in the near future.

Appendix 8
The postcode system

1   The postcode is a combination of up to seven alphabetical and numerical characters. These define four different levels of spatial unit and are used to facilitate automatic sorting and delivery of mail by the Post Office throughout the United Kingdom. The postcode system was designed by the Post Office and remains under its control.

2   There are 1.5 million postcodes covering approximately 22 million addresses in the United Kingdom. Some addresses (around 170,000) have their own unique postcode because they receive large quantities of mail per day. These are the Large User Postcodes. The remaining addresses are covered by the Small User Postcodes and on average there are 15 addresses to each small user unit postcode although this number may vary considerably depending on the nature of the local postal service, eg. in urban or rural areas. In a sample of 400 unit postcodes in Sheffield the distribution of the numbers of households in each unit postcode had a wide variation with about 4% having more than 65 households each and a few with over 100 households.

3   The operation of the postcode system is best explained through an example. *(See Figure 1 overleaf)*

MK42 9WA

Each postcode is structured in two parts, the first is the outward code **(MK42)** the second the inward code **(9WA)**. In this example **MK** identifies the Postcode *Area,* each Area comprises an important unit in the national transport network. In this example **MK** refers to Milton Keynes. Currently there are 120 Postcode Areas. Each Postcode Area is divided into smaller geographic units called *Districts* which are referenced by the numerical figure in the first half of the postcode (**42** in this example). There are 2,700 Postcode Districts in the United Kingdom. Postcode Districts are divided into smaller geographic units called *Sectors*. These are designated by the figure which begins the second part of the postcode. In this instance the figure is **9**. There are 8,900 Postcode Sectors. Finally the complete unit postcode pinpoints one street, or part of a street, with the last two letter characters, in this case **WA.**

4   Approximatley 18,000 changes per year are made at the unit postcode level over the whole of the country as a result of new development or demolition of buildings or if the postcodes available for a particular district have been used up.

The Postcode Address File (PAF)
5   The PAF is a computer file containing all addresses in the country. Names are only included on the file if they are necessary to identify an address. The addresses are held in postcode order and are separated into:

   1. Large Users of the post ie. the 170,000 users who, because they receive large quantities of mail per day, have their own postcode. These users are specified by name of organisation and, although large users are generally larger businesses, a proportion will be made up of other types of organisations such as hospitals, schools and Government Departments.

   2. Small Users of the post ie. the remaining 22 million addresses. Small user addresses are mainly households, but a proportion is made up of business and other non-residential addresses.

6   The PAF can be used to supply addresses for geographic areas defined by postcodes. These checklists can be supplied with or without a 'premises code' or PREMCODE, which is an additional code given to addresses. It is made up of the first four characters of the address, thus the PREMCODE for 31 Baker Street is 31BA, and when used in conjunction with the unit postcode provides a means of uniquely identifying every Small User address in the country.

7   The PAF is used commercially to match addresses with postcodes and for a variety of other purposes. It is maintained by the Post Office and updated every 3 months from information supplied by Post Office postcode teams based in each of the 120 postcode areas. Variations in the type of information supplied by these teams have led to some inconsistencies in the PAF.

Central Postcode Directory (CDP)
8   This Directory contains the Ordnance Survey Grid Reference (to 100 m resolution in England and Wales and to 10m resolution in Scotland) and local government ward code for the first address in each postcode. The whole of Great Britain is covered. The Directory is maintained by the Office of Population Censuses and Surveys (OPCS) and the General Register Office (Scotland) (GRO(S)).

9   The link between postcodes and grid references forms a powerful system which can be used for a range of locational analyses, route planning, calculating and checking journey times, locating remote addresses, and ensuring that depots are located at the most convenient and economic points. The link between postcodes and ward codes enables a user to link the statistics published by OPCS and GRO(S), derived from Census of Population data, to the Postcode Sectors, Districts and Areas.

Postcode boundaries
10   Postcode boundaries are available on a variety of commercially produced maps. Postcode Sector boundaries are available on 1:100,000 scale maps and Postcode District boundaries on quarter inch to the mile maps, the index map for these products shows the Postcode Area boundaries. In addition three inch to the mile conurbation maps showing Postcode

Districts and Postcode Sectors are available commercially. The Regional Newspaper Advertising Bureau has funded the digitisation of the Postcode Sector boundaries.

11 The GRO(S) used a postcode based system in the planning of Enumeration Districts for the 1981 Census of Population in Scotland. Unit postcode boundaries were mapped on Ordnance Survey base maps (at various scales, but mainly 1:2,500 in built up areas). The resulting Enumeration Districts consisted of sets of unit postcodes, thus enabling the linkage of Census of Population data to other datasets having unit postcode references to be carried out.

Appendix 8 Figure 1 **Operation of the postcode**

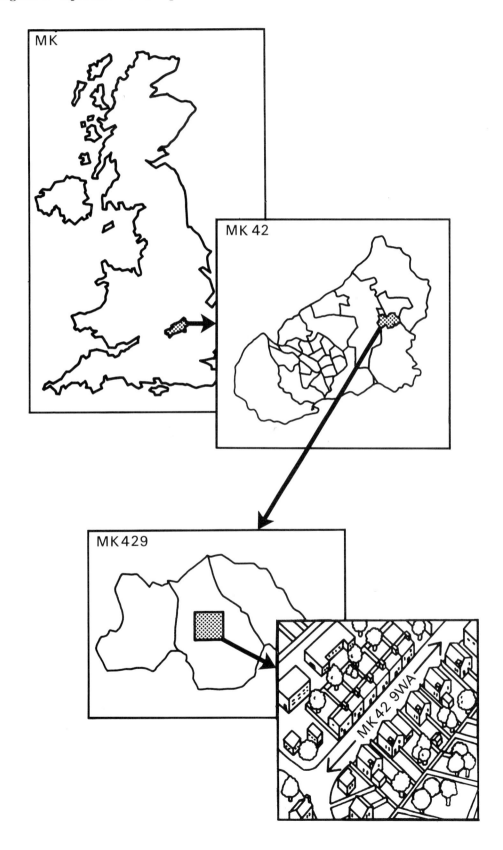

Appendix 9
Selected bibliography

The following references have been selected from the 5000 or so publications known to have been produced on Geographic Information Systems and related areas. Those shown below have been picked to cover a wider range of topics from theory to the operational system, including consideration of standards, data quality and user education. Wherever possible, examples have been selected from easily available sources: the bulk of the references are thus drawn from the Proceedings of the Auto Carto London conference, held in September 1986. The two volumes of these proceedings, edited by M.J. Blakemore, are available from the Royal Institution of Chartered Surveyors.

ANDERSSON S. (1986): The Swedish Land Data Bank *Proc. Auto Carto London*. Vol.2, 122-8.

BURISA: Quarterly Newsletter of the *British Urban and Regional Information Systems Association*. Frequent articles on Geographic Information Systems applications.

BURROUGH P.A. (1986): *Principles of Geographic Information Systems for Land Resources Assessment*. Monographs on Soil and Resources Surveys No.12, Clarendon Press, Oxford.

BOUILLE F. (1986): Interfacing Cartographic Knowledge Structures and Robotics. *Proc. Auto Carto London* Vol. 1,563-71

BRAND M.J.D. (1986): The Foundation of a Geographical Information System for Northern Ireland. *Proc. Auto Carto London* Vol.2, 4-9.

CATLOW D.R. (1986): The Multi-disciplinary Applications of DEMS (Digital Elevation Models). *Proc. Auto Carto London* Vol.1, 447-55

CHRISMAN N.R. (1986): Obtaining Information on Quality of Digital Data. *Proc. Auto Carto London* Vol.1, 350-8.

DANGERMOND J. (1986) GIS Trends and Experiences. *Proc. 2nd. International Symposium on Spatial Data Handling* 1-4. *Seattle*. International Geographical Union

DENEGRE J, DESCHAMPS J.C. and GALTIER B. (1986) Landsat and SPOT High Resolution Images: a New Component for Geographic Databases. *Proc. Auto Carto London* Vol.1, 527-37

DIAZ B.M. (1986) Tesseral Addressing and Arithmetic — Overview. *Spatial Data Processing Using Tesseral Methods*, 1-10. NERC, Swindon

ENGLAND J.R., HUDSON K.I., MASTERS R.J., POWELL K.S., and SHORTRIDGE J.D. (1985) *Information Systems for Policy Planning in Local Government*. Longman

GILFOYLE I. and CHALLEN D. (1986) Digital Mapping — the Cheshire Experience. *Proc. Auto Carto London*, Vol.2, 472-9.

GOODCHILD M.F. and RIZZO B. (1986) Performance Evaluation and Workload Estimation for GIS. *Proc. 2nd. International Symposium on Spatial Data Handling, Seattle* 497-509. International Geographical Union.

GREEN N.G., FINCH S., and WIGGINS J. (1985). The 'State of the Art' in Geographic Information Systems. *Area* 17, 4, 295-301.

GREEN N.G. and RHIND D.W. (1986). Teach yourself Geographic Information Systems: the Design, Creation and Use of Demonstrators and Tutors. *Proc. Auto Carto London*, Vol.2 327-39

JACKSON M.J. and MASON D.C. (1986). The Development of Integrated Geo-Information Systems. *International Journal of Remote Sensing*, 7, 6, 723-40.

McMASTER P., HAYWOOD P.E. and SOWTON M. (1986). Digital Mapping at Ordnance Survey. *Proc. Auto Carto London*, Vol.1, 13-23.

MOELLERING H. (1986) Developing Digital Cartographic Data Standards for the United States. *Proc. Auto Carto London*, Vol.1. 312-22

OPENSHAW S. and MOUNSEY H.M. (1986) Geographic Information Systems and the BBC's Domesday Interactive Videodisk. *Proc. Auto Carto London*, Vol.2, 539-46.

OS (1986) *The National Transfer Format*. Ordnance Survey, Southampton.

SAMET H., SCHAFFER C.A., NELSON R.C., HUANG Y-U., FUJIMURA K. and ROSENFELD A. (1986) Recent Developments in Quadtree-based Geographic Information Systems. *Proc. 2nd. International Seminar on Spatial Data Handling, Seattle* 15-32, International Geographical Union.

SMITH P.M. (1986) Digital Mapping in Land Registration and its Relevance to the Possible Development of a Geographic Information System in the UK. *Proc. Auto Carto London*, Vol.2, 129-38

WIEBE V.N. (1986) The Burnaby Experience with Computerized Mapping. *Proc. Auto Carto London*, Vol.2, 102-11

WIGGINS J.C., HARTLEY R.P., HIGGINS M.J., and WHITTAKER R.J. (1986). Computing Aspects of a Large Geographic Information System for the European Community. *Proc. Auto Carto London*, Vol.2, 28-43

WILLIAMSON I.P. (1986) Trends in Land Information Systems Administration in Australia. *Proc. Auto Carto London*, Vol.2, 149-61

WOODSFORD P.A. (1986) Cartographic Digitising — Technical Trends and Economic Factors. *Proc. Auto Carto London*, Vol.2, 489-95

Appendix 10
A selection from the written evidence

The Committee was greatly assisted in its work by the wealth of excellent and thought-provoking evidence submitted to it.

In our view, the quality and volume of the evidence reflects the importance which many attach to the subject of our Enquiry and we are grateful to all concerned for the care and effort taken in preparing responses to our questions. We recommend that the specialist reader of our Report should consult the complete record of evidence which will shortly be deposited at the Department of the Environment Library.

For the general reader, however, we felt that it might be helpful if this selection from the written evidence were published with our Report. It demonstrates the very wide range of user, practitioner and researcher who contributed so helpfully to our study; it also reflects the diversity of views expressed to us, with not all of which, by the nature of things, were we always able to agree.

This is no more than a selection, however, and space has not allowed us to print much other evidence which has been of value to us. We hope that witnesses whose evidence has not been included will understand our difficulty and will find their views expressed in the words of others whom we have quoted.

INDEX

A   Geographical information

ORDNANCE SURVEY
The Committee of Enquiry is taking evidence at a time of rapid technological development and many people think that as a result sophisticated GIS are just around the corner. Almost anything is seen to be possible, and those who do not automatically support this rosy view are considered to be lacking in vision. Ordnance Survey has been capturing digital data for over 10 years and has found that it is a complex operation. It is against that background therefore that Ordnance Survey tempers its enthusiasm with caution. However, let there be no doubt that Ordnance Survey does believe that digital mapping

and GIS have a long term future and that Ordnance Survey will continue to work towards their successful development.

SCOTTISH OFFICE
While developments in remote sensing of physical characteristics will be of interest to us and monitored by us for their potential usefulness, it is likely that the area of greater advance will be in the consolidation and improvement of existing geographic information and in the establishment of the best and most cost-effective way of using it for policy purposes. One particular field where the pace of interest has been quickening recently is that concerning land use and environment, this pace is expected to continue; and it may be one where by the end of the 5 to 10 year period mentioned in the Committee's invitation, the impact of remote sensing will be considerable.

INLAND REVENUE (VALUATION OFFICE)
The introduction of Information Technology is expected to improve in a major way access to the database and to provide opportunities for analysis and manipulation of information denied to us under the present manual system. The falling unit cost of both storage and processing facilities will enable the cost effective application of Information Technology to improve office efficiency and provide scope to perform new functions. The overall aim is to improve the valuation and statistical services we provide to client departments and authorities but we also expect to make more efficient use of scarce and expensive staff resources.

NATIONAL REMOTE SENSING CENTRE
The present situation regarding spatial information technology is a paradoxical one. There is a natural unwillingness on the part of those organisations that we see to be potential users of geographical information systems to commit money or manpower either to investigate the uses of such systems within their organisation or to train members of their staff in the relevant techniques until the capabilities of geographical information systems have become more clearly apparent and the costs more easily estimated. Similarly, data-collecting organisations such as the Ordnance Survey are cautious in committing themselves to the large-scale provision of data before they have been convinced that such provision would be commercially viable.

ECONOMIC AND SOCIAL RESEARCH COUNCIL
ESRC sees a particular need to consider geographic information on a wide-ranging basis, with the prospect of relating environmental data on topography, land use, climate and environmental quality with social, economic and demographic data. This means organising, indexing and integrating existing data sources — both from remote sensing, field surveys and Ordnance Survey and from censuses, manpower information, service data from central and local government and surveys from public and private bodies — so that data on specific sub-national

spatial units can be made available to researchers.

BRITISH LIBRARY
Irrespective of the specialist non-current databases which may become available at particular research nodes, it seems that the concept of a series of basic data sets collected over time should be discussed and such a series be considered as a central obligation and funded accordingly.

THAMES WATER AUTHORITY
A useful example is the work of the Planning Section in the Technical Services Group of Rivers Division. The Thames engineer is required to consider the implications on Thames Water's interests of a proposed new development. He will need to consult one area, and then mark up the new proposals on the map. He must then consult a second set of maps to determine the flooding information in the area, a third set of maps for definitive information regarding 'main rivers' adjoining or crossing the site, a fourth set of maps to gain up-to-date topographical information; then he may have to consult other sets of maps on trunk sewers, water mains, boreholes, current improvement schemes, etc. It is simply not feasible for all this information to be shown on one set of maps, and there are never the cartographic resources fully to update the maps. However, digital mapping appears to offer a long awaited opportunity for viewing all this information.

EDINBURGH CITY COUNCIL
It is felt that the quality of decision-making could be improved if the spatial element of the various information bases were to be subjected to more thorough and systematic analysis. The relatively superficial analyses undertaken at present might be attributed to a number of factors, including the shortage of trained staff as well as shortcomings in the data itself. Some information is simply not available in a spatially referenced format at anything more localised than the city scale; some information is available at the very local level (eg by address or postcode unit) but not in a format which can be readily translated or aggregated into meaningful areal units without a great deal of work; and other information is available for manageable areal units, although these units are rarely consistent between different data sources (eg wards, postcode sectors, health board districts, job centre areas, parliamentary constituencies) and are often subject to boundary changes.

SOUTH OXFORDSHIRE DISTRICT COUNCIL
Apart from the NJUG experiment, a House of Lords enquiry and a few other lack lustre looks at parts of the problem, nothing has been done with regard to INFORMATION AS A RESOURCE, its uses and its spatial nature.

This examination of information as a resource and its uses is essential to any discussion on spatial links. It is the very exercise which identifies the duplication and the savings and/or efficiencies which can be achieved by the adoption of a computerised mapping system and its associated alpha-numeric database.

GRAMPIAN REGIONAL COUNCIL
This authority has already devoted significant resources to the consideration of the use of remotely sensed data for planning purposes. It was realised that this development work involved considerable commitment which perhaps could not be justified on the basis of the restricted overall resources available to the Council in the short term although the potential value of such a system was considerable. Based on this experience it is considered unrealistic to expect one local authority to take the initiative in investing in the development of a system which could be of national significance in terms of geographic information. It is suggested that the government should have at least a pump priming role in financing trial applications of such systems.

CHESHIRE COUNTY COUNCIL
Given appropriate mechanisms for managing, transmitting and analysing geographic data easily and cheaply, the use of such data by County Council departments for planning and operational purposes seems likely to grow. A number of future developments have already been identified in Cheshire. These include:

1   Aggregating information on housing development, school catchment areas, public transport and land in County ownership to give a context for school planning in the Education service.

2   Aggregating demographic information with data on the capacity of homes for the elderly to provide a basis for planning future provision for the elderly.

3   Aggregating demographic data with records of criminal activity and police beat areas to assess where changes in the latter are required.

4   Automatic generation of optimum delivery routes for County vehicles from a digitised route network plus data on collection and delivery points.

5   Using information on the location of farms and livestock types to quickly define quarantine zones in the event of an outbreak of a notifiable disease.

6   Identifying library catchment areas to assist in planning of libraries and allocation of mobile libraries in rural areas.

7   Resource planning for the Fire Service to ensure adequate fire cover and response times for the County using urban and fire risk type data.

8   Digitised schedules of road maintenance programmes and location of street furniture.

INTERNATIONAL COMPUTERS LTD
It is ICL's belief that the geographic context of data underpins much of the work of both public and private sector organisations and this context provides a powerful mechanism to analyse the data and inform decision making. However, this application area is not yet mature and hence there is great deal of confusion about the way ahead and the potential benefits.

The present situation is one of fragmentation, in practically every dimension that one cares to examine. While it seems to be generally accepted that geographic information systems are an important future application area, there seems no coherence in the present activities.

This fragmentation can be seen in user requirements that are not clearly understood and therefore have produced no consensus on system strategies to provide solutions; in the plethora of 'pilots' and 'trials' that have taken place and are still planned; in the way data, from the many data sources available, is not produced with geographic information system use in mind; in the split between the advocates of raster technologies and the advocates of vector technology; in the way cartographic and thematic maps have little connection; in the concentration on graphics and the relative neglect of data management and networking; in the different perceptions and financial circumstances of different segments of the market; in the lack of standards.

INTERGRAPH LTD
The development of computers will continue to provide ever cheaper processing power and data storage capacity which will enable the user to get faster access to larger databases, and allow the software developer to make use of improved progamming techniques which will result in programs which will do things we cannot even conceive today.

In addition the development in:

Networking/Communications
Scanning
Display Techniques

will enable GIS to grow by faster input of data, remote access to central databases and better update of that data.

ASDA STORES LTD
Main types of information used are:

1981 Census data

OPCS Ward maps

Family Expenditure Survey data on spending commodity group, by region and by socio-economic group.

Wherever available, Local Authority data detailing population change after 1981 and forecast population change.

Details of food competitors throughout the country (size, location, parking etc). Source: Unit for Retail Planning Information, Institute of Grocery Distribution plus internal Asda intelligence gathering.

Details of the floorspace allocated to various non-food categories of shopping within existing shopping centres Source: CACI Market Analysis.

County Structure Plan and District/Local Plans for information on current planning policies.

TESCO STORES LTD
It is anticipated that over the next 5-10 years the type of spatially referenced information required by the Research Unit will remain fairly constant. Thus the main categories of geographic data will continue to be:

Census of Population
Other Government Sources
Digitised Road Network
Branch Data
Competition Data
Customer Survey

Internally collected data is continuously updated while data from government sources can only be updated periodically. The Census of Population provides the greatest problems in terms of the currency of data and it is felt to be of top priority that a quinquennial Census of Population be introduced.

ROYAL INSTITUTE OF BRITISH ARCHITECTS
The basic information needs of private consultants and those employed in the public sector will be the same, although varying in depth. They can be identified as:

(a) geographic information concerning the layout and use of

land, its features, levels and contours, degree of exposure to the natural elements, geological structure and natural water courses.

(b) information concerning man made services, underground and overhead, which can be identified as:

(i)   water mains

(ii)   sewers

(iii)   land drains

(iv)   electricity services

(v)   gas services

(vi)   pipelines − oil and gas

(vii)   telecommunication services

(c) information concerning

(i)   underground railways

(ii)   road tunnels

(iii)   aircraft flight paths

ROYAL SOCIETY
This submission interprets the term 'geographic information' to mean any information related to the Earth's surface, the matter beneath it, or the enveloping atmosphere or hydrosphere that is, or can be, spatially referenced with respect to that surface. Such information is necessary to all scientists concerned with the terrestrial environment and to others concerned with the representation and measurement of the Earth's surface.

The handling of geographic information is a topic that cuts across both departmental and disciplinary boundaries and, for that reason, tends to be neglected as not being central to the interests of any one agency or discipline. In a period of economy in public expenditure, particularly in the collection of data and the funding of teaching and research in higher education, this tendency is exaggerated. Yet technological developments in computing, affecting the storage and cost of handling data, and in remote sensing, which has the potential vastly to increase the amount of data available, make it increasingly important to look across departmental boundaries.

The importance of monitoring changes in the global environment has been increasingly recognized in recent years, and British scientists and governmental agencies have an important contribution to make here. Such changes, for example in mean sea level, may have direct consequences for the United Kingdom. ICSU, which embraces all sciences of interest to the Royal Society, held the first symposium on global change in Ottawa in September 1984 and has established a steering committee to further that work. The International Geographic Union and the International Cartographic Association have a joint working party examining the feasibility of establishing a world digital database for environmental data. Among international agencies, the United Nations Environmental Programme is developing a Global Environment Monitoring System (GEMS) and has a related project for a Global Resource Information Database (GRID), and the Food and Agricultural Organization has taken an active interest in the development of GIS within its field of responsibility.

ROYAL TOWN PLANNING INSTITUTE
Over the next 5-10 years, planners are likely to be increasingly concerned with:

(1) Emphasis on active promotion of development in addition to consideration of development control proposals.

(2) The management of land as a resource (e.g. as a means of achieving policy in other areas such as housing and economic development).

(3) The planning and management of resources in rural areas.

(4) Local economic development.

(5) The reduction in the gap in living conditions between deprived areas and the rest of the community.

(6) The call to provide information to the public on problems and proposals.

(7) The need to co-ordinate proposals (particularly in the metropolitan areas) with neighbouring local authorities and other public and private agencies with development interests.

The needs of planners for spatial information over the next 5-10 years therefore fall into the following major categories:

(1) The current picture of the spatial world in which planners operate (with maps being fundamental as a highly convenient model of reality indicating the location of human and physical features).

(2) Land and its use/condition/potential use (including agriculture, forestry and vacant/derelict land, and with particular emphasis on land for housing and economic development).

(3) Buildings and their use/condition/potential use (with particular emphasis on residential, commercial and industrial buildings).

(4) Information about networks including those related to highways, transportation and public utilities.

(5) Population and its characteristics (including employment, social and economic deprivation, and interactions with land/buildings such as shopping habits and journey to work).

(6) Information on neighbouring areas (for categories (1) to (4) above) in terms of consistent and compatible information describing the current situation and policies/proposals.

ROYAL GEOGRAPHICAL SOCIETY
The Society gave evidence to both the Ordnance Survey Review Committee in 1978 and the House of Lords Select Committee on Science and Technology − Remote Sensing and Digital Mapping in 1983 and much of what follows is consistent with that evidence. Since these enquiries the advance in information system technology has continued rapidly but its development and expoitation in geographic information systems in the United Kingdom has slipped during the lengthy consideration of the reports of these enquiries. Thus much of this evidence is necessarily concerned with geographic information systems. Throughout this evidence what are described as geographic information systems are those which contain any information which can be spatially referenced.

ROYAL INSTITUTION OF CHARTERED SURVEYORS
As a general comment, the commercial user group apears the most dynamic and best informed. The level of sophistication and understanding is probably the highest as most commercial users have a clear understanding of their requirement and have already developed and implemented systems to meet that requirement. Commercial expediency justified the funding.

Members' requirements and priorities for datasets vary with their professional interests. All believe that Ordnance Survey large scale digital data must be derived from the common data-base. Thereafter there is a need for datasets such as:

(i)   All underground utilities

(ii)   Planning data

(iii)   Property ownership and boundaries

(iv)   Land use

(v)   Land values and property valuation

(vi)   Mineral workings

(vii)   Socio-economic data

KINGSTON POLYTECHNIC
It is readily apparent that there is a considerable potential market for geographically-referenced data in many walks of life ranging from commercial activity to the pursuit of leisure. Although text-based information systems are now in relative abundance in small businesses, spatial information systems are by contrast rare, despite the relevance of geographic data to many applications. For example, 'High Street' estate agents possess information systems to enable them to maintain records of clients, accounts, and addresses, but few, if any, possess systems to enable them to instantly generate maps showing the locations of all properties of a specified type, their positions in relation to vital local amenities and the local topography despite the obvious importance of this information to clients. The need for such systems is clear and the hardware required to generate it is comparatively low cost and often already posessed by the companies concerned. The major reasons for lack of such systems appear to be a poor level of awareness of the value of spatial data in the community, and the absence of suitable datasets and software for their manipulation.

BRIGHTON POLYTECHNIC, COUNTRYSIDE RESEARCH UNIT
However, if this Research Unit's experience is in any way an indicator, it would be fair to suggest that the demand for integrated spatial information systems is only just beginning to gain momentum in Britain, and is some ten years behind the position in the United States of America. if the current piecemeal activities continue in Britain, considerable public investment is likely to be placed in the autonomous collection of generally local data sets of county size and much smaller, virtually all of which are incompatible for a variety of reasons.

UNIVERSITY OF BRISTOL, SCHOOL FOR ADVANCED URBAN STUDIES
We would emphasize that the use of geographic information is likely to become more rather than less important in future years. The coming decade is likely to see further changes in economic and social processes and structures, and the spatial consequences of these changes for local (regional, urban or neighbourhood) policy are likely to be complex. Changing relationships between work and home, changes in the patterns of family or household behaviour, technological and/or administrative innovation influencing the spatial provision of services would all place a premium on the availability of spatially sensitive data for policy and administrative purposes.

BIRKBECK COLLEGE, UNIVERSITY OF LONDON
Progress in the development of expert systems and knowledge-based GIS should also greatly increase the scope and speed with which information can be handled. Additionally, the integration of high resolution remotely sensed data may become

of great importance when used in conjunction with many other kinds of geographic data.

CSIRO, CANBERRA AUSTRALIA

The term 'land-related' has been given its broadest meaning, to encompass information about natural resources, the environment, land ownership, land use, transport, communications, mapping, demography and socio-economic factors, in fact any information that can be related to position. The information is said to have a 'spatial component'. This component may be identified by any number of means, such as local government area, electorate, postcode, town, land parcel, line, network, point or grid cell.

Generally, it is information about features or activities on the earth's surface, but can include sub-surface, aquatic and atmospheric phenomena. Examples of land-related data are:

administrative (land ownership, valuation and rating, zoning, improvements to land, taxes, political unit boundaries),

cadastral,

infrastructure (transport, communication and utility service networks),

cultural (demography, education, health and other socio-economic factors),

industrial (manufacturing, trade, building and construction),

environmental and natural resources (vegetation cover, topography, soils, water, ecology, climate and agriculture).

US DEPARTMENT OF THE INTERIOR, BUREAU OF LAND MANAGEMENT

A GIS generally comprises integrated hardware and software and is used to capture, store, retrieve, edit, manipulate, analyze, and display spatial data. Computer-based GIS's were originally conceived to solve problems associated with storing, manipulating, and displaying very large volumes of geographically referenced data in conventional map and text form. The number of different cartographic and earth-science data types that could be analyzed was limited by the user's ability to mentally integrate several types and produce new maps from selected combinations of data sets. GIS technology permits cost-effective analysis of the many cartographic and earth-science data types using economic and scientific models that consider the many factors needed to make policy decisions, develop resource management plans, or make resource assessments and appraisals. However, to describe a GIS as a system for handling 'map-based' information is convenient but misleading. GIS technology can be used to manipulate spatial information and data whether derived from a map or any other source.

The need for GIS technology within the Department of the Interior, is based on at least five trends over the last two decades:

(1) Increasing numbers of large cartographic and earth-science data bases have become available.

(2) Remotely sensed data have become available in digital form and are often combined with cartographic data.

(3) Data manipulations have become more complex and quantitative models are being developed to assess the relationships between various resource parameters.

(4) Rapidly emerging computer technology developments have provided more efficient ways to store, retrieve, and manipulate geographically referenced data.

(5) Natural resource and environmental issues have gained increased scientific and political attention, demanding that new technology be developed to provide more rigorous, systematic means to evaluate environmental conditions and assess potential impacts.

B Digital mapping

HM LAND REGISTRY

HMLR, in common with other OS map users, is also a little concerned about the likely rate of progress of the OS digital mapping programme, and whether the programme of OS digital map conversion will be in step with HMLR's possible future requirements. HMLR looks to OS to produce the digital maps which it may require in the future, but it must be said that if the results of HMLR's own feasibility study point to the possibility of securing significant economic and operational benefits from using digital mapping, it may find it necessary to consider seriously the feasibility of using scanning techniques for converting OS maps where neither OS or any of its official agencies is willing or able to provide the service. This is not a practicable proposition at the present time, because of the relatively high cost and not entirely satisfactory results obtained from scanning, but improvements in both these areas seem possible in the future.

Although a sophisticated feature coding facility exists, the data is held in a very unstructured format, and causes problems for users such as HMLR, who have to devise their own procedures for complex manipulations of the data. There appears to be a need to review the structure of the data, and to bring it into line with current user needs.

Other problems which arise as a result of the way in which OS digital data is prepared and presented, relate to the positioning of text and matching of detail at sheet edges. As to the latter, it is recognised that this problem is inherent in the method of production of the original large scale map documents themselves, and digitisation perpetuates, and sometimes magnifies an existing problem. It is known that automatic edge matching software is available, but it could give rise to gross errors in the hands of an unskilled user. Ideally, future editions of digital OS maps should be edge matched at the production stage. In some cases this might imply field visits to check detail.

If HMLR were to proceed to the design and installation of a computerised land registration mapping system, it would require a pattern of digital mapping coverage which would enable it to introduce the system on phased basis, by the conversion of its map records relating to land in successive local authority districts. To some extent, HMLR may be somewhat different from other users, who may, for example, regard the digitisation of urban areas within Greater London and the Metropolitan Counties as the main priority. HMLR is potentially the major customer for OS large scale digital maps, as ultimately it is concerned with the boundaries of all land parcels throughout England and Wales. In the circumstances, it is suggested that HMLR's particular requirements for digital map coverage need to be given full weight when the planning of a programme to extend digital map coverage is considered.

ORDNANCE SURVEY

The dangers of duplication of data capture are a concern to Ordnance Survey. The current digital mapping programme is being accelerated both in-house and by placing contracts in the private sector, but Ordnance Survey is aware that it cannot satisfy all its potential customers as quickly as either Ordnance Survey or the customers would like. However, standardisation is an important step in the orderly development of GIS and Ordnance Survey views with disquiet the waste of resources incurred by users who commission independent digitising of Ordnance Survey maps to a different digital specification than that used by Ordnance Survey itself.

It is easy to amalgamate the number of codes but it is impossible to subdivide individual codes without recoding. Many feature codes in digital mapping are included for generalisation purposes and while it is true to say that the generalisation problem has not been solved yet, it is likely that improvements in data processing technology will eventually overcome the problem. It would be unrealistic to redigitise later solely to capture generalisation information.

Any question of demand must be tempered by OS views on the feasibility of scanning Master Survey Drawings (MSDs). A recent consultancy has indicated that it is not possible to scan MSDs because their function as working drawings creates problems in the scanning process. Suggestions have been made as to how to avoid using the MSD as input to the scanner but these have still to be tested. The OS master negatives have stipple incorporated to indicate the roof areas and when scanned this stipple introduces problems during the conversion from raster to vector data. Thus there is no simple way in which scanning can be used without further investigation. At present the quality of the input documents, the sensitivity of the scanners and the software to convert from raster to vector, all combine to limit progress.

That having been said, OS has evidence that a few users are interested in background mapping produced as a raster facsimile. There is greater interest in the concept of vectorised raster scanned data with minimal feature coding − but rarely for identical sets of feature codes. Similarly, there is some interest in vectorised raster scanned data with geometric correction.

Manual methods of capturing existing topographic data at Ordnance Survey have probably been optimised and current projections indicate that the capture of basic scales mapping information by such methods and with present levels of resources will not be completed until 2015. Of course, it is almost certain that technological advances will speed this caputre and raster scanning − allied perhaps, to some form of pattern recognition − presently seems to offer the best hope for an earlier completion data, but although the resolution of scanners is acceptable now, the software to convert scanned data into a form suitable for manipulation by GIS is not yet good enough, particularly for handling descriptive information about the data. After discussions with commercial developers, Ordnance Survey think a suitable raster based system may be available in about 5 years time.

Improved speeds of data capture are a vital requirement for both Ordnance Survey and its users, and Ordnance Survey believes that the only realistic way this can be achieved is through technological development. Raster technology is moving in the right direction, but there are still problems to be overcome, particularly over the incorporation of information describing the data that has been captured and its relationship to other adjacent or remote data. Some work has been done on pattern recognition, but further research is required.

Data structures developed in the last 10 years are likely to form the basis for software trials in the next 5-10 years, although there is known to be research into the use of hardware or firmware to solve data processing problems that are currently handled by software. As software solutions are significantly slower than either hardware or firmware solutions, the use of the latter 2 methods will improve data access and increase the speed of data manipulation. Additionally these solutions will lead to the evolution of less complex storage structures than previously thought necessary, while still allowing complicated database relationships to be maintained.

Obviously data structure is closely linked to data processing. Parallel or array processing coupled with developments in file scanning, is likely to have a major impact on GIS development, as it will increase the speed of processing and the ability to search through very large datasets. Ordnance Survey is aware that trials using an array processor have shown potential for great improvements in some aspects of spatial data processing, using either raster data or a combination of raster and vector. The recently announced development of the transputer is likely to lead to the widespread availability of parallel processors.

Hardware developments, however, usually require associated software development if the true potential of the new technology is to be fulfilled. Computer languages may soon be able to handle efficiently the vaguely defined requests for information that may be the hallmark of GIS users. If these language developments are combined with knowledge based systems then very powerful programs could be made available. Considerable work along these lines is being undertaken outside the traditional cartographic and geographic disciplines, such research should not be ignored.

FORESTRY COMMISSION
The digitising of OS 1:10,000 maps for rural areas would certainly act as a tremendous catalyst toward further interest and development. However, the Forestry Commission would need to be convinced of the benefits and proven capabilities of systems before their acquisition. Our managers' expertise is naturally in forestry and not mapping, and therefore any systems would need to be simple to use and flexible enough to respond to the particular problems of forest management.

MINISTRY OF DEFENCE, MILITARY SURVEY
The 1985 amendment to the Copyright Act in respect of computer software does not appear to take account of data capture and storage other than in relation to literary work. Early clarification of the copyright issue is essential.

ORDNANCE SURVEY
All Ordnance Survey material is Crown Copyright and this copyright extends to maps made by others but based on Ordnance Survey material. Ordnance Survey authorisation is required before copies can be made.

Computer handling of maps will extend the range of use, but this will not affect the rights of copyright owners which are protected through the Copyright (Computer Software) Amendment Act 1985. This is important to Ordnance Survey as copyright protection is essential if the creation and updating of "works" are to be financed in this way.

NATIONAL JOINT UTILITIES GROUP
Utilities will have commented individually on their copyright arrangements. The NJUG view is that charging must be realistic

and cognisant of the fact that the utilities are captive customers. It is also important that Ordnance Survey recognise the large amount of revenue they receive from the utilities and take this into account in formulating their marketing and development programmes. Similarly, if the price of digital tapes increase beyond a reasonable price, then market forces will intervene forcing utilities to other suppliers or to produce the maps themselves. If that is unco-ordinated the benefits of the common background map will be lost.

COUNTY PLANNING OFFICERS' SOCIETY
The most pressing copyright issue concerns Ordnance Survey mapping. The Ordnance Survey obtains much of its revenue from copyright fees which are currently assessed on the basis of the number of individual copies made of their ordnance maps and plans. Once OS data is widely available in digital form, suitably equipped users will be able to use it, with other data, to produce maps, thus reducing the need for copying in the traditional sense. OS will encounter difficult problems in determining how much OS material has been copied and in devising a pricing structure which realistically reflects both the value to users (and in particular to local authorities) and the costs incurred by OS in producing suitable data. Clearly, copyright on geographical information needs to be clear.

WESSEX WATER
Current copyright restrictions concerning the use of OS maps act as a major constraint in the liberalisation of the digital mapping market. It is felt unlikely that significant progress will be made until restrictions are removed and different organisation allowed to compete and provide a range of services more closely matched to the diverse needs of the various users.

ROYAL INSTITUTION OF CHARTERED SURVEYORS
In general our members feel that the present system of Ordnance Survey copyright for large scale maps works reasonably well and that charges are about right. However, it is recognised that even the present system of copyright for graphic products is difficult, it not imposssible, to police, and that there are apparent anomalies in copyright charges to similar users.

As users move increasingly into digital data, policing any system of copyright will become even harder. Indeed, we believe that widespread use of digital data must throw into question the whole concept of copyright and copyright charges on large scale plans.

We can see the need for the OS to make reasonable charges for supplying data in the first instance; and increasingly, of course, as digital updating becomes possible, it will be only the updating which is supplied to regular users. Beyond this we are inclined to think that, assuming the initial purchase price reflects reasonable usage of the data, copyright charges should be dropped.

ROYAL SOCIETY
Copyright and charging present major problems. The ideal, from a scientific perspective, would be for all publicly collected data to be in the public domain, as in the United States, where the use of such data is not constrained in any way and where charging is confined to the cost of handling the data. We recognize that governments in this country have adopted a different stance, and public agencies have been instructed to maximize revenue from copyright rather than to maximise use of the data they collect, but we contend that this stance is detrimental both to science and the economy. In 1978 in its comments on the report of the Whitford Committee on

Copyright and Design Law, the Society remarked that there was considerable confusion between the use of information contained on a map and reproduction of the map or portion of the map. This confusion is exacerbated when the information is in the form of digital data.

ROYAL GEOGRAPHICAL SOCIETY
The level of charging for the supply of geographic information should be one that maximises its use.The present system of an increasing search for copyright income has the reverse effect.

ROYAL TOWN PLANNING INSTITUTE
The Institute views with concern the possible effects of copyright on the interchange of geographic information. The Institute considers that advances of information technology and the need for the rapid exchange of geographic information in the future, merit serious reconsideration of how the Copyright Act of 1956 applies to digital mapping data.

CABLE TELEVISION ASSOCIATION
In general, the Cable industry needs access to up-date large scale digital maps at a reasonable cost. There also needs to be progress towards a suitable method of interchanges of information betweeen the cable industry, utilities and local authorities, without the need to use hard copy.

It may be possible to provide these facilities in a less detailed way than that currently proposed by Ordnance Survey. We believe the speed at which digitisation of the mapping records is carried out in this country needs to be speeded up substantially. One way of doing this might be a more flexible approach by Ordnance Survey towards outside assistance from others able to carry out the work. If cable is to reach the majority of the country in the next 15-20 years, it is imperative that decisions are taken quickly on an accelerated programme of map digitisation, as well as a method of acquiring, producing, providing and exchanging such information. Unless this is done, it will not be possible for the cable industry to benefit from such techniques when installing cable systems.

THAMES WATER AUTHORITY
There is an urgent need to plan for the national programme for digitising maps to be compressed into 3 years, with the cost possibly shared by Government departments, public utilities, Local Authorities and other principal beneficiaries, in order to avoid enormous duplication.

BRITISH TELECOM
The task of providing a digital map background for the UK, with the necessary standards of accuracy and detail must fall to Ordnance Survey. When this is available and the time is right, individual businesses can, if they wish, decide to digitise their own spatially-related information on graphics systems of their own choice.

BRITISH GAS CORPORATION
A major difficulty identifed by the Utilities is the absence of a comprehensive set of Ordnance Survey maps on digital tape. Small areas of the country have been digitised by O.S., but only limited coverage is presently available in the urban areas where most Utilities plant is buried to coincide with customer locations. At current digitisation rates based on their current digitisation techniques, O.S. consider that full digitisation will take a further 30 years. O.S. currently digitise their maps using over 120 levels of data. Utilities require less than 10 levels of data from the existing maps and have little need for the high extremes of accuracy constantly striven for by O.S.

Digital tape costs from O.S. are due to increase substantially. The present cost of £30 per tape, is projected to rise to £150 over the next few years. This is of particular concern when the amount of data on tape far exceeds the Utilities needs.

The Utilities can employ digital tapes on a wide scale only when:

(a) they become available in sufficient numbers to cover Regions or operating areas within regions, and

(b) the Corporation has digitised its own plant records against appropriate coordinates following investment in the necessary hardware, software and training etc.

There are undoubted savings and operational benefits to accrue from the adoption of computerised information systems by the Utilities. Some of the benefits that accrue from the availability of O.S. digital maps have been demonstrated fully in a joint trial with NJUG/OS and the highway authorities at Dudley. This trial was described in the Corporations Part A Evidence. Because of this success so far and the great potential for further development the O.S. digitisation timescale of 30/50 years is totally unacceptable. The Corporation and other NJUG members agree that 5 to 7 years would be nearer the extremes of acceptable timescales.

The O.S. meet the Utilities twice each year to exchange views and the Utilities have expressed the above concern on many occasions in the past and continue to do so. In view of the extended unacceptable timescales envisaged by O.S., the Utilities are actively pursuing alternatives to shorten the timescales and hasten technological change.

One of the principal constraints is the lack of OS map information in a form suitable for modern electronic office systems, with little prospect of it being available for several decades in a form suitable for the Utilities.

Reference was made, in the response to question 5, to the excessive amounts of detail in the O.S. digital tapes. Some 120 levels of data are to be digitised and this far exceeds the needs of the Utilities by a factor 10 or more. In addition the completion date of 2015 is too protracted for business purposes. The Utilities would be prepared to trade off excessive levels of data for adequate levels of data to meet Utility needs within more realistic time scales of 5 to 7 years.

NATIONAL JOINT UTILITIES GROUP
If utilities, local authorities and other public organisations were working to the same background map, it would be unnecessary duplication if each of them were to hold separately the identical information. For example, on the basis that all individual utility information was spatially related to the same background map, only utility based information need to transmitted for information exchange. This would require access to a common background map, kept current by a central agency.

Those users that are already embarking on digital mapping cannot wait 30 years for the background maps to become available. They are therefore digitising the maps themselves, to meet their own requirements. Where these digitised maps do not meet the requirements of other users covering the same geographical area, it is likely that when they require the background map, they will do their own digitising. The consequence of this would be:

(i) duplication of effort;

(ii) duplication of data storage;

(ii) incompatibility of background map.

As demonstrated in the Dudley trial, another important benefit from linking and sharing data accrues from the local authority adding to map backgrounds proposals for new developments. This would avoid the duplication of effort by each utility adding this information to its own plant records. The data of the new development, as in Dudley, would need to be held separately from the remaining map data so that the final positions could be archived when construction is complete and the buildings have been surveyed.

SOUTH WESTERN ELECTRICITY BOARD
During the course of the development of the Board's procedures for transfer of map records to a digital base a close liaison has been established with officers from Ordnance Survey. At the time when the Board's research began OS had already produced maps for selected areas of the country in digital form and were seeking the views of potential users on their specification for these maps. The argument centred around the need to separately identify common items held on OS maps, such as road and footpaths lines, house outlines, boundaries, contours, spot heights etc. It was the Board's strongly held view then, and this opinion has not been changed in the light of our subsequent experience, that the OS proposal to separate data into more than 100 feature codes is an over complication. It seems inevitable that this complexity delays progress when maps are digitised manually and increases the resistance to the introduction of automatic computer based map scanning techniques.

It is apparent to the Board, however, that the development of automatic digital scanning systems such as that used by SysScan (UK) Ltd could make considerable savings in the time required to transfer records maps from their existing form to the computer based form. At the present stage of development the scanning process can take a good quality OS map background and create initially a raster image of this which is then transposed by software into a vector form. At the same time limited feature coding separates the pecked lines of footpaths and text from other map features. Inevitably this process is not perfect since character recognition by software can never be as well resolved as is achieved with the understanding applied by cartographic draughtsmen. It will be readily understood therefore that to be able to feature code automatically the lines of cables separate from other lines drawn on maps will be difficult. A possible way forward is that a scan image is made of the map background, and a second scan image of the record map. Subtraction of the 2 images by the computer would expose the additional features added by the Board when the record was created. An editing process could them be applied to the exposed single level record to create the division into feature codes. Such a process could be likened to time lapsed photography of the heavens to identify the movements of planets etc.

WESSEX WATER
It is accepted that the constraints on Ordnance Survey are such that digitisation of the entire country may take up to 30 years, and that steps are currently being taken to carry out some work externally, as a means of accelerating the process. Indeed, the use of outside bodies to assist in the preparation of digital maps should be encouraged. However, DM technology would appear to offer at least a partial solution to this problem in the existence of a number of feature codes representing 'levels' of the map.

The digitisation of a subset of feature codes could be carried out in appreciably less time than complete digitisation; salient data (for example road and property boundaries) could thus be made available to users more immediately than would otherwise be possible. The order of priority for feature codes could be determined by consultation with major users, as could geographical priorities. This approach may be preferable to the injection of large amounts of public money which would otherwise be necessary.

GRAMPIAN REGIONAL COUNCIL
Digital mapping will be of fundamental importance in determining the ability of the Council to make better use of the geographic information it needs to plan and run it services. This is true of general needs for maps, the addition of other data to digitised maps (especially, for example, public utility data), and the linking of mapped information to property databases. Three factors will determine the ability of the Council to take up the opportunities offered by digital mapping − the need for better data from digital mapping; a quicker completion by the Ordnance Survey of its national digitising programme; and the development of the relevant software for Scottish users.

SOUTH OXFORDSHIRE DISTRICT COUNCIL
The main areas of duplication are in the numbers of Ordnance Survey maps in use, of both map-based (graphic) and associated alpha numeric information and finally in the ways the information is used − the task is duplicated. These 3 areas of duplication apply to all organisations internally but the position nationally is much, much worse perpetuating similar duplication between organisations. The result is a huge waste of monetary and manpower resources with added costs incurred in correcting erroneous acts arising from inadequate or inaccurate information − there can only be one BEST SOURCE!

INTERNATIONAL COMPUTERS LTD
Underlying much of the spatial context of data sets in the UK is of course the output of the Ordnance Survey. If the possibilities of modern information technology are to be exploited, the provision of digital map data is crucial. Access to this data from OS is only possible if one is interested in the small minority of the country that is covered by OS digital maps. Potential customers in other areas are forced to consider alternatives, which inevitably require a cost/quality trade-off decision to be made. As many of the alternatives produce lower quality and higher cost alternatives than OS digital data, there would seem to be grounds for OS to reconsider its strategy, both in terms of data quality standards and price. For example, OS could decide to quickly provide a lower quality coverage, which could be incrementally refined at increasing price.

The OS digital data structure has not really been designed with use in sophisticated computer based systems in mind. It is more appropriate for an alternative transmission medium for traditional paper maps.

SYSSCAN UK LTD
Test results so far encourage SysScan to confidently predict that by applying an artificial intelligence system to OS large scale data from the Kartoscan range of scanners a vectorised map product can be produced in a relatively short time with about 20 accurately assigned feature codes. Possible features for recognition include, significantly, buildings, land parcels, roads, railways, footpaths and water features.

SysScan believes that a national policy of scanning the large scale map base of the UK and selling this data in raster form would have the following advantages:

- full national coverage cheaply;
- full national coverage quickly (5 years?);
- accuracy of all mapping will conform to OS standards;
- control of base map supply will remain with OS;
- raster data will provide input to vectorising and coding production lines to produce more conventional digital products;
- raster data will be suitable for field update and production of raster digital SUSI's;
- raster data will be suitable for 'print on demand' production of hard copy via laser and matrix printers.

INTERGRAPH LTD
Although many utilities at the present moment require less accuracy and quantity in terms of feature codes than is currently being provided by the OS this should not influence the production of the 1:1250 digital maps. It is much easier to work from the whole to the part; it is virtually impossible to work the other way without a major degradation of the national asset carefully preserved over the last 100 years or so.

ROYAL SOCIETY
The Ordnance Survey, as the national mapping agency for Great Britain, is central to the development of geographic information systems in this country. The standards that it adopts, the data that it provides and its policies for making those data available will have a profound influence upon the pace and manner of future developments. It is critically important that the standards adopted by the Ordnance Survey meet the whole specturm of needs and not just the requirements for producing the traditional map, and that its policies for marketing its digital data should be reconsidered so as to encourage more use of those data. Without such use and exploitation of a national resource the ultimate commerical benefits of digital data will be poorly realized and this country will fall behind its international competitors.

It must be recognised that, for most organisations, the acquisition of OS data is only one step on the way to adopting digital data handling. They also have considerable investment to make in hardware, software and in digitising their own records. High OS charges could prevent many from setting out along this road.

We believe it would be a mistake if present Government funding for OS digitising is seen as 'priming-finance' or a loan to be recovered against sales of digital data. Digitising the National Archive should be seen as a once-off capital cost which is unrecoverable.

A futher cause of duplication is the very long time scale envisaged for completion of the digitizing of OS maps and the high standards the OS requires; many users, such as institutes of the Natural Environment Research Council, cannot wait that long and may be satisfied with digitizing to lower standards (and hence at lower cost). Such digitizing is often poorly documented and there is no comprehensive register of what has been digitized; in any case, there are deterrents to the exchange of such data. In so far as duplication exists, it could be minimised by greater openness about the availability of data and a less restrictive attitude on copyright aimed at maximizing use. We believe that an attempt should be made to create a comprehensive register of what has been digitized, although we realize that this will be difficult until the problem posed by copyright are resolved. It is in the national interest to minimize duplication and to remove obstacles to the exchange of digital data.

ROYAL INSTITUTION OF CHARTERED SURVEYORS

There should be some re-assessment of the content specification for OS large scale maps. On the one hand, is it really necessary for the OS to show 'minor detail' in backgardens when it seems to be so little used? On the other, should OS maps not show at least some skeleton of street furniture, which would act as a framework, to which utilities information could be tied, in spatial data systems?

Ordnance Survey maps are a multi-purpose topographic data base which is put to a wide variety of applications by different users. This being so, accuracy standards must be maintained.

THE STANDING COMMITTEE OF PROFESSIONAL MAP USERS

The rate of up date and the availability of digital data is too slow at present to avoid a real risk of alternative digitising sources being developed. We think that the rate of digitising should first be accelerated, and thereafter carefully maintained on a long term basis.

Digital bases have the capacity to give selective information. We believe the first need is to make such selective information available to office and private users on a large range of scales and by methods which are quick and convenient in terms of accessibility and costs.

COUNTY PLANNING OFFICERS SOCIETY

There is one major sphere in which central government has a key role and that is in the funding of the Ordnance Survey and its operations to ensure adequate and speedy provision of the digitised 'geographic base'. Without in any way prejudicing the requirement that the Ordnance Survey be as profitable as possible, the injection of funds to ensure geographic cover in digitised map and plan form at 1/1,250, 1/2,500, 1/10,000 and 1/50,000 scales is a prerequisite if this task is to be undertaken within a realistic time scale. At current levels of funding and technology it is understood that the OS 'best estimate' is that it will take some 30 years to complete a programme of digitising the 1/1,250 and 1/2,500 scale plan cover.

LOCAL AUTHORITIES ORDNANCE SURVEY COMMITTEE

In this section LAOSC would like to refer particularly to the House of Lords Select Committee on Science and Technology Remote Sensing and Digital Mapping – Recommendation 30. This stated: 'The Department of the Environment in collaboration with others should fund a software project to make Ordnance Survey digital data more usable by and useful to local authorities. In their response, the Government recognised that it is important that the software through which Ordnance Survey digital data can be put to use by local authorities should be up to date and efficient. Local authorities are especially concerned as at present there is a clear divergence between local authorities (user need) and Ordnance Survey (data supplier) in the field of digital mapping.'

ROYAL TOWN PLANNING INSTITUTE

Digital mapping is an urgent and fundamental requirement. The current barriers to the expansion of digital mapping need to be removed – in particular early national coverage needs to be ensured and an 'Ordnance Survey recommended' software package developed which will cover map data base management, flexible map plotting and the restructing of map data for use in information systems. Facilities offered by digital mapping which are of importance to planning include:

- The ability to select characteristics from a wide range in order to focus on particular problems.

- The ability to produce maps at a wide variety of scales across convenient map junctions.

- The potential for storing additional data as overlays for superimposition on the Ordnance Survey base (for example proposed road lines, location of public utilities etc.).

- The ability to compile land and property boundaries for incorporation in information systems.

- The speed and efficiency of manipulation, minimising expensive time in map preparation.

Positive encouragement of the use of digital mapping must also be actively pursued, for example through holding down the cost of digital data and subsidising joint experiments between Ordnance Survey and users to gain two-way experience of user needs and the potential of digital mapping.

BIRKBECK COLLEGE, UNIVERSITY OF LONDON

One of the most obvious examples of duplication is that of digital topographic data produced by the OS. Much *ad hoc* digitising of these data – at both large and small scales – has been done by other institutions and it seems likely that this situation will continue until the OS can provide a complete coverage of data at a competitive price.

C Availability of information

MANPOWER SERVICES COMMISSION

Where Government action is needed is in the area of data security and confidentiality. There are circumstances whereby Official Statistics for industrial employment are confidential under the 1947 Statistics of Trade Act, due to assurances included in the questionnaire forms, whereas commercial agencies are now able to duplicate the same data without need for such assurances. The effects on the operation of NOMIS are felt at two levels, firstly that of employment figures for individual firms, ERI data, and that of summary data, ERII. Although most users of NOMIS in central government and county councils have separate legal access to ERI data the University staff of NOMIS itself are not permitted to be authorised users and cannot receive or mount the data.

INLAND REVENUE (VALUATION OFFICE)

We have mentioned the demand for disclosure of information in respect of transactions in land and buildings. If this was thought by Government to be advisable in the interests of improving information available to the property market there would appear to be no further objection to including transaction evidence in a national register of property information.

Provided problems of confidentiality can be overcome the advantages of a national register of property information to commercial and other users would, we believe, be very great. There would also be benefits to conveyancers. The Conveyancing Committee in its second report under the

Chairmanship of Professor Farrand (presented to the Lord Chancellor in January 1985) has recommended, in paragraph 9.15, that

'... The immediate need is for the computerisation of information pertinent to searches and enquiries of HM Land Registry and local authorities. In the long term, comprehensive property databases using unique property reference numbers and digitised maps, should be created. For this purpose, the production by the Ordnance Survey of digitised maps for the whole country should be expedited.'

We believe there is unsatisfied demand for statistical and other analysed property information. While it is not considered appropriate for the Valuation Office to set up computer systems with the primary function of meeting the needs of other users we must anticipate outside demands for the supply of information held primarily for Valuation Office purposes. Progress in meeting the demands of other users will depend in part upon decisions affecting the disclosure of transaction evidence.

ORDNANCE SURVEY
A central Government commitment to make its own data available for incorporation into major GIS is a pre-requisite for the long term benefits of GIS to be realised.

SCOTTISH OFFICE
We doubt if one comprehensive database (or 'bumper box') is the way to proceed. All datasets have their peculiarities and their reliability and value in analysis can degrade substantially when removed from the charge of those directly involved in their production and use. Even assuming the economic and technical feasibility of linking geographic datasets, it is still to be expected that much of the use of any one dataset will be within its own sphere, with little or no combination with data from elsewhere. This may point to the maintenance of primary database designed to meet the needs of principal users. Faster means of data transmission between agencies should, however, facilitate greater linking than at present.

COUNTRYSIDE COMMISSION FOR SCOTLAND
In the immediate future, the Commission expects to continue to have need for the same wide range of data on rural resources and on recreation as described earlier, namely topographic data, survey data for many aspects of the natural environment and data about recreational resources and recreational activities. One priority area for development in the future is to seek information about change to the countryside, which is a subject of considerable current interest to a large number of organisations. Data about landholding in Scotland would also be of considerable value to the Commission and although it may be unfashionable to say so, conventional maps and aerial photographs will continue to serve many of the Commission's needs for day-to-day information of a geographical character. We welcome, of course, the fact that developments such as the digitising of Ordnance Survey maps provides a means of rapid updating of map content, although − for topographic maps such facilities inevitably better serve urban areas, where the main effort in survey updating is placed, than the countryside.

Debate about geographical information systems carries a presumption that the computing power involved in these systems can readily manipulate data on the basis of complete coverage for an area under study. It is important to recognise that, for many topics of interest to bodies having responsibilities relating to the rural environment, data at a reasonably fine grain are not available across the nation in a uniform and reasonably current form. Thus, there is no map of land use distribution for Scotland; no current map of semi-natural vegetation categories; some areas of Scotland await revision of one inch geological maps last published in the previous century; the coverage of soil and land capability maps at the one inch scale is limited; and even the 1:10,000 Ordnance Survey mapping of rural areas has a significant proportion of maps that have not been revised within a decade.

The Commission appreciate that new sources of data, such as satellite imagery, will eventually enable the mapping of attributes for which no nationwide data exists at present, but it is our understanding that interpretation of such imagery requires continuing development work before this kind of remote sensing can be considered appropriate for routine use in Britain. The Commission supports further research expenditure in this field.

ECONOMIC & SOCIAL RESEARCH COUNCIL
There are major inconsistencies in charging policies for government data which are not easy to understand, in particular the very high cost of Census data to the academic sector (albeit provided at a generous discount) as compared with most other government data. ESRC would maintain that where data has been collected with the aid of public funds publicly-funded research ought to be granted concessionary rates as a general principle. The level at which charges for data are in general now set are such that many academic institutions, with their now straitened budgets, are unable to afford them and the consequential loss in potential research of value to the country at large is not to be underestimated. The cost of purchasing data, moreover, is at the direct expense of funds required for the support of research itself.

BRITISH ACADEMY
The Academy is also concernd about the archiving of both analogue and digital data, and although it is aware that the latter is being considered by the British Library, it believes that too narrow a view may be taken of this topic, especially of the need to ensure that such data are held in formats that permit different categories of data to be related to each other. The Academy is also concerned at the apparent lack of interest of many official agencies in preserving such data once their immediate usefulness for official puposes is past. Even from the perspective of the official organisations themselves, such data will be essential for establishing trends and measuring change. The Academy is also concerned that machine-readable data may be lost by not being properly archived, i.e. by neglect rather than by deliberate destruction. The Academy sees a need both for consultative machinery to ensure that digital archival data are not destroyed before the academic community and other potential users have the opportunity to comment on their usefulness for academic and other uses, and for clear national policies on the archiving of such data. It is appreciated that the storage of data has a cost, but so too does their conversion into digital form. The costs of storage are falling rapidly in real terms whereas conversion of analogue data into machine-readable form is still a lengthy process which severely restricts much potential academic use of such data.

NATIONAL JOINT UTILITIES GROUP
If the inquirer has access to a visual display unit he could have access to all utility records. Safeguards against misuse have to be built in. In Dudley this has been achieved to a certain extent by the 'Private' and 'Public' level concept.

WATER AUTHORITIES ASSOCIATION

In the longer term, the Water Authorities will be seriously examining the opportunities offered by remote sensing, particularly in its applications to environmental protection of coasts and rivers, flood warnings, rainfall forecasting and surveying, and the use of satellites for communications purposes. The Water Authorities are becoming major users of rainfall data supplied by the Meteorological office from radar sites. A possible national standard for such data is under consideration by the Water Authorities. Concerning the dissemination of satellite data, there is evidence of duplication of effort which could be reduced by the development of a national strategy. The Water Authorities would welcome an opportunity to participate in the consideration of a national strategy to meet the wider user needs.

DEPARTMENT OF THE ENVIRONMENT

In general, the Department's view is that there is not a substantial degree of unnecessary duplication in the collection and management of geographic information systems. There are instances where several users might wish to maintain separate data holdings but this can be justified in usage terms.

HM LAND REGISTRY

As regards the areas of undesirable duplication in the collection and management of spatially referenced information, if this were widespread, it would form an important part of a case for the justification for the linking or sharing of data sets. In practice, there does not seem to be a great deal of evidence to support the view that widespread duplication of effort exists.

NORTHERN IRELAND OFFICE

One area of duplication that we are aware of is the detection of new housing, independently by OSNI, by those concerned with property valuation, by the rating authority, by those maintaining the electoral register, and by others. The consolation for the multiple costs incurred is that the detection processes are thought to be independent; and so cross-checking the results of the independent exercises (which is done by all or most of the organisations mentioned) allows a statistical estimate to be made of the completeness achieved.

COUNTY SURVEYORS SOCIETY

Problems of duplication occur where different non-compatible systems operate within different areas of the same department, where collection and plotting systems are not compatible with design systems, and where the same information is held by separate authorities such as district and county councils. The greatest problem in using spatially referenced information supplied by other organisations is posed by the incompatibility and proliferation of software for reference systems.

LAMSAC

Data is often collected more than once because the individual collectors will not share or divulge it due to confidentiality or restricted use constraints. This can happen within local authorities and between central government and other organisations and local authorities. In some cases the restriction on use is statutory eg. "It can only be used for the reason it was collected" but in others it can be the result of individual attitudes.

TESCO STORES LTD

The major areas of duplication in the collection and management of spatially referenced data relate to information which is not readily available from one central source. That is, data which is collected internally or at the local government level. There is inevitably a lot of duplication in the collection of such data but whether or not this is considered desirable depends on the point of view from which it is considered. This duplication of effort might not be desirable from the point of view of those ultimately providing the finance (ie the customer or the rate payer), but it might be considered desirable by a commercial organisation that wishes to maintain a competitive edge over its major rivals.

SOUTH CAMBRIDGE DISTRICT COUNCIL

All duplication in this context is undesirable and often organisations are ignorant of information held even by similar organisations. A central "catalogue" of information held would be a first step in enabling information to be shared. Whether that information would be freely available would depend upon the policy of the Information Provider towards charging.

EDINBURGH CITY COUNCIL

One of the most intractable problems raised by the present debate is that of confidentiality. This is a particularly sensitive issue in view of the proposal to pinpoint information as accurately as possible by means of grid references. However, the treatment of the Small Area Statistics in the 1981 Census has demonstrated that it is not insurmountable. Provided that the information required is not too deeply cross-referenced a postcode base would generally offer the best compromise between locational accuracy and anonymity. Special precautions would have to be taken so that 'large-users' with individual postcodes are not identified. As a further guarantee of privacy, it may be necessary to appoint a data security officer, who would be responsible for ensuring that information is only released at an aggregated level.

ASSOCIATION OF DISTRICT COUNCILS

Difficulties in interpretation arise when processing statistical information when the levels of aggregation required are at parish or district-wide level. In some cases, eg employment data which is compiled by Employment Service Area, it is simply not possible to aggregate the information into a form meaningful to a non-metropolitan district. These problems could be resolved by the adoption of standard units for all information, eg enumeration districts, with software available to enable users to aggregate the information into other units more suitable for their own purposes

DERBYSHIRE COUNTY COUNCIL

Generally data should be referenced at the lowest spatial unit possible, compatible with confidentiality, and the reference should be a grid reference or postcode. When it is necessary to go to large spatial bases for statistical summary or for confidentiality reasons Wards should be used.

With a common referencing base and modern communications between computers there is no reason why data bases cannot be held centrally (either by Authority or Region) with sub-stations having direct access. Most data would be discreet to the collecting Authority and if a national library of types of data held was kept, then there should be no problem in potential users identifying and seeking permission to use that data. Again the key to any national or regional referencing system is in the base geographical locational reference and I would reiterate that this should be OS grid reference based.

GRAMPIAN REGIONAL COUNCIL

Many local authority departments require to provide data for a wide range of aggregated areas and as such the requirement is for data at the lowest level of disaggregation allied with the facility for automatic aggregation. Often disaggregated data is held to be confidential but it needs to be recognised that the data would be used and presented only in a series of reaggregated forms. Interest in data at the individual or highly disaggregated level is only a means to an end in itself and is of no consequence to satisfying many service requirements. Restrictions should therefore relate to the lowest aggregation level acceptable for publication.

One of the principal concerns of the Council is that much of the information which is required by local authorities is collected indirectly by Central government and other agencies, for example, population counts from Primary Care Records, floorspace statistics from Assessor or Inland Revenue, migration and property address data from Electricity Board records. Local Authorities are often forced into using proxy methods of acquiring or collecting this data which is wasteful of resources and may not provide an adequate alternative. Central government must accept a role for Local Government in deciding priorities for data bases, their content and format, and must make data available to Local Government.

HILLIER PARKER MAY AND ROWDEN

The District Valuer feels unable to make data available below district level for confidentiality reasons. We believe it is unnecessary for the rateable value, floorspace and use of a property to be confidential, and hope that the Committee will recommend that the legal constraints imposed on the District Valuer be lifted, and that the Valuation Service be encouraged to make information more readily available.

Much is talked of the coming major changes in information technology, and how easy this will make the accessing of large amounts of data. Unfortunately, in this particular major field. the evidence suggests that we are moving backwards and not forwards. We hope that the Committee will address itself with some urgency to the problem of the production of local area data, in a standardised form which is consistent over time.

What is needed is readily available information at local level covering matters such as demographics, economics and the stock of commercial floorspace, and how that has varied over time.

Whilst it is possible to obtain all this local information through undertaking detailed studies, it is not feasible to spend a great deal of time when deciding whether or not to bid in the market place. It is therefore of great importance to have available standardised information which is consistent over time.

There has been an alarming trend that the published information on which we have relied has decreased. Firstly, the Census of Distribution, which gave information on the retail sales in town centres, was stopped and no information is available later than 1971. Secondly, the Department of the Environment has announced its intention to cease publishing Commercial and Industrial Floorspace Statistics.

ASDA STORES LTD

Whilst Asda remains primarily a retailing organisation we will continue to require spatial data on population size, character and spending patterns for small areas. At present our forecasting approach involves merging data from OPCS sources and from the Family Expenditure Survey. Problems occur mainly with differences in definitions between these sources. To be of maximum benefit to a retailer such as Asda, we would like to see FES data at a smaller area than regional — preferably sub-divisions of counties.

We would like to have the 1991 Census information produced in a spatial format eg. grid squares that can be loaded directly into our computer.

A new Census of Distribution similar to that of 1971 is needed.

HEALEY & BAKER

Our comments on the availability of statistical information relate first to sources relating to retail development, and second to office and industrial development.

So far as retail development is concerned we are particularly conscious of the lack of up to date and detailed information relating to shopping on a nationwide basis. The 1971 Census of Distribution is now seriously out of date, especially in towns and cities where major development has taken place in the last 15 years, and the cancellation of a Census of Distribution in 1981 means that there is now no recent and authoritative source of information about shopping centres generally to which reference can be made.

Although the Retail Enquiries have attempted to fill this gap, even the 1977 Enquiry which dealt with a limited sample of towns is of relatively little assistance and unless steps are taken to remedy this deficiency in the near future, it seems to us that the quality of the basic information upon which important decisions relating to shopping are made will become even poorer.

So far as Healey & Baker are concerned this has 2 important consequences. First, in the investment field, it is increasingly difficult to advise clients on recent trends in shopping growth and likely future developments. Reference back to the 1971 Census is no longer realistic and, therefore, many investment decisions are being based upon totally inadequate and out dated information.

So far as town planning advice is concerned similar considerations apply in relation to the advice given to clients on the development of new shopping facilities. In many areas local authorities do not even have an up to date floorspace survey, and there is, of course, virtually no information as to turnover since the 1971 Census. This lack of an adequate and independent statistical base can lead to unnecessary disputes and uncertainty in the planning of shopping and in our view the only way in which these deficiencies can be remedied is by a full Census of Distribution to the same level of detail carried out in 1971.

The second main field of activity relates to office and industrial property, for which, again, there is only limited information available and one of the potential sources of data, the Commercial and Industrial Floorspace Statistics publications are, we understand, to be discontinued.

TESCO STORES LTD

Increased accessibility to the raw data is another area in which central government could initiate development. Data could thus be made available for users to do their own tabulation and analysis, a situation which does exist at present with regard to data from the General Household Survey.

There is a need for all these bodies to undertake R & D in this field. Government departments are the obvious bodies for developments relating to Census variables and other such data sets, especially in the collection and referencing of data. Government departments could also increase accessibility to data sets and encourage users to do their own tabulation of the data.

LOCAL AUTHORITIES ORDNANCE SURVEY COMMITTEE

A concern must be that much of the information required by local authorities is collected indirectly by central government and other agencies, and local authorities are then forced into using proxy methods of acquiring or collecting this data which is wasteful of resources, and may not prove an adequate alternative. Often, Government departments when collecting information do not take into account the use of such information to local authorities or the saving in time and effort which might be made if additional information was collected at the same time. There is also the possibility of paying twice for the same information from one external source, unless a central catalogue of information held was compiled. Duplication of the collection, handling and management of information between local authorities has alwasy existed, but is being reduced. An aspect which must not be forgotten is duplication by local authorities themselves, where because of practice or confidentiality data is not shared, and has to be recollected in each case. This is disadvantageous and wasteful of resources.

Central Government departments and agencies are major suppliers and collectors of information. Therefore Government must realise that it is not a unilateral body, but has a co-ordinating and instigating role to play in conjunction with local authorities. It would help local authorities if they standardised the location systems under which data is supplied, standardising levels of charging and simplifying confidentiality and copyright rules. It is unlikely that one single organisation charged with compiling and maintaining a nationwide data base could either keep the information up to date or meet the demands of users. However, a centralised library of information could be effective. It would need agreed standards for location referencing and units of aggregation, etc. Such a library could also include information regarding software supplies. There is scope for an exchange of information between Government and local authorities, and this should increase for the future. In this context a very useful liaison committee was the Co-ordinating Committee on Locational Referencing, and it has been greatly missed.

Information needs to be accurate, complete and up to date in the interests of both users and providers. Local plans, however, are prepared within the constraints of reasonably available information as many local authorities do not usually have the resources (manpower or financial) to undertake extensive research or purchase outside information. It is also possible that local authorities might consider the purchase of Ordnance Survey digital map data if they were available sooner than currently planned, and were less expensive, at a slight cost to accuracy.

HOUSE BUILDERS FEDERATION

Data on key variables such as migration, employment and unemployment, earning and earnings data are not available at sufficiently disaggregated levels to facilitate the effective analysis

of market or policy areas in the field of housing. Indeed problems of data disaggregation are therefore unlikely to be confined to housing and must crucially affect the forward planning of employment, retailing, transport and other factors.

Key areas of land use planning and other future service provision could be greatly improved by the establishment of a system of common building blocks of data. In theory it should be possible to establish a 'basic spatial unit', for example the enumeration district which could become a common reference point for all demographic economic and social data. This basic building block could then be developed through district, county and regional level to provide a common hierarchical database which could then be used in the programming and provision of a wide range of facilities and services.

HBF is concerned that the quality of geographical data generally available is being significantly eroded by cuts in public expenditure and by changes to the local government system. There has been a noticeable reduction in local authority monitoring outputs over the past few years and the quality of overall forward planning must inevitably suffer from this. The abolition of the Metropolitan counties will seriously impair the collection of data at metropolitan county level and the Federation shares the concern of many public and private sector bodies that the quality of forward planning and service provision in the metropolitan areas will suffer as a result of these problems.

ROYAL SOCIETY

On-line access to spatially referenced data is generally not necessarily in scientific work, and the large quantities of data that would need to be transmitted would make transfer over public telephone links both slow and costly. However, a more pressing need is for on-line access to information about information, e.g. dates, times, flight paths and cloud conditions, on Landsat imagery of the United Kingdom held at the National Remote Sensing Centre, the availability and characteristics of aerial photography of the country, and the dates and coverages of maps. We welcome the intention of the Ordnance Survey to establish a computer-based catalogue of its maps and possibly of information on aerial photography. We regard these as essential and urge that all such indexes be accessible to scientists over the public telephone work.

REMOTE SENSING SOCIETY

Remotely-sensed data have been used for many decades in scientific research in the Earth and Atmospheric sciences, and in management and planning of the environment. In the past the main source of such data has been aerial surveys, which have been carried out from time to time by the Royal Air Force, or commissioned by local authorities. Aerial photographs are still the most widely-used type of remotely-sensed data for practical management. In scientific research, and in certain applications such as weather forecasting, digital data acquired mainly by satellites are now in widespread use. The role of aerial photographs should not be overlooked when considering data sources; such data are used routinely by many agencies throughout the United Kingdom, for example in town and country planning, forestry and environmental management.

ROYAL TOWN PLANNING INSTITUTE

Central government collects and manipulates many important data sets, some of which are made available to local authorities. Those that are currently made available, for example the Census of Employment and the JUVOS statistics are supplied with different geographical references. Data that is currently difficult to obtain from government sources includes the detailed floorspace statistics collected by the Inland Revenue. These statistics would be particularly useful on a shopping centre base – rather than on a local authority base, as they are published nationally. Hopefully, some progress will be seen on this soon when the Inland Revenue implements its computer based property records system.

BRIGHTON POLYTECHNIC, COUNTRYSIDE RESEARCH UNIT

A National Data Base Directory containing details of spatial referencing, cell structure, methods of data capture and data sources, copyright, ownership and availability (on-line or magnetic tape) is seen to be the first essential step which would bring users into contact and lay the foundations for reducing the amount of duplicated effort which is already taking place. On-line access would be likely to be the most efficient method of sharing or transferring data but the use of magnetic tapes would also be effective even though more time consuming. The feasibility of establishing central facilities for generating compatible tapes for use on a variety of main frame computers would be necessary. This will be particularly important in international data transfers.

Access to development control, housebuilding completions, rating files, property registers and location of registered agricultural holdings would all be valuable for *authorised* users *subject to the constraints* of the Agricultural Census Act, Data Protection Act and copyright. It is not recommended that this information should be readily available for general research but that it should be restricted to existing users or those delegated officially by them to carry out work on their behalf. The delegated or commissioned user would then be subject legally to the same standards of data protection as for exmple the central or local government department.

D   Linking data

DEPARTMENT OF THE ENVIRONMENT

The Department sees some benefits in the ability to link relevant data sets when informing policy issues. Such linking should enhance the value of data through extending the scope for bringing together cost-effectively data from different but relevant sources. Duplication of holdings should be reduced because of ready access to the authoritative holding. Custodians of such linked data sets, being under wider spotlights, might be encouraged to improve and maintain more accurate and up-to-date information.

In general the Department has experienced no major difficulties in handling geographic information systems or sub-sets provided by others. However not all systems have consistent areal bases and the Department has had to resort to the use of 'best fit' algorithms in some cases when using information from some other sources. On the whole these algorithms are regarded by the Department as sufficiently robust for the purposes for which the data are used. A recent example is the 'best fit' algorithm generated to translate wards in parish based Rural Development Areas, thus enabling a wider range of economic variables to be included in analysis for policy considerations.

ORDNANCE SURVEY

The ability to link data sets is the essence of GIS. If GIS are to justify their costs, they must facilitate faster and more accurate

decision making by allowing the interpretation and analysis of closely related data sets that previous manual systems just could not combine. The savings in time by not having to duplicate data capture, and the benefits of accuracy through having the experts in any particular set of information being responsible for its capture and maintenance of that information are important. It is equally important that common standards apply to the format and classification of data holdings and that any deviations from that standard are widely publicised.

For GIS to work effectively, collectors of geographic information must be prepared to make their data available to others, either on-line or by some other means. In a utopian world, all GIS users would have unlimited on-line access by means of user friendly software to a wealth of information in the public and private sector, for a nominal fee. Utopia is however unlikely for a variety of reasons, and while anything that leads in that direction can only be good, users will have to be prepared to pay for access.

A major initial constraint to the integration of many data sets will be the differing levels of accuracy and precision in the individual data sets, which could lead to some people being inhibited in the use of the data. It is important that users are aware of the potential applications and limitations of particular data sets.

One of the important means of linking different data sets lies with the access key. While the user, for instance, may enter a system through names and addresses or land parcel numbers, the programmer will have to convert this information to a file key. Ordnance Survey argue strongly that the key should be National Grid co-ordinates rather than any other system. The National Grid is highly flexible giving access to units of 100 kilometres square or unique millimetre locations with equal ease. All other building blocks such as boundaries, postcodes etc can be related to the National Grid. The National Grid is immovable whereas postcodes and boundaries for instance are liable to change, and when they do, a master file linking them to the national grid co-ordinates has to be amended.

SCOTTISH OFFICE

Quite apart from identifying the threshhold of justification or specification of a linked database, there is likely to be a difficulty in ensuring compatible spatial references for different sets of data arising from different sources. Within the Scottish Office this is likely to be considerable, since there is interest in linking a range of socio-economic material with physical material. Whereas the physical material is increasingly grid referenced and should in principle be capable of linkage, most of the socio-economic material is not so referenced. Where its reference is reasonably detailed, it is generally by means of postcode. To date we have limited experience of linking postcode referenced material with grid referenced material – and even where the postcode reference is at the most disaggregated level (ie postcode units) some difficulty seems likely. Digitisation is the key technique which facilitates point and zone spatial relationship to be established for analysis and presentation.

NORTHERN IRELAND OFFICE

Sharing datasets – as we plan to do via an eventually integrated, distributed database – leads to enormous potential benefits. It speeds up, and reduces the cost of, those operations where, at present, data is requested, sought, extracted (if found), transmitted, and used clerically. Such data sharing is very frequent in the public sector for a variety of reasons ranging from statutory obligations, down to sensible applied economics.

There are problems in growing such a system around a nucleus. The major one is that organisations with insufficient vision are still defining single, independent systems, or working slowly towards (full or partial) internal integration. They apply sound enough principles in their work − for instance, the dictum from CCTA that one should find the software that best meets the requirements, and buy the hardware that supports it. But they do so on the scale of their current preoccupations, and not on the scale which is now becoming possible. thus, when presented with the larger possibilities, they ask questions like 'what will compatability with this network cost me?', and refuse to contemplate questions like 'What will incompatability with this network cost me?'. They want assurances that all of the organisations whose data they might wish to access should be signed up before they will even contemplate participation, never mind signing up to reassure the organisations who might wish to access their data.

INLAND REVENUE (VALUATION OFFICE)
We believe that a national system is needed for the precise identification of property units to facilitate the linking of data sets and the exchange of information. We consider that a national system should be based on the Ordnance Survey National Grid, although there will be a need for the Valuation Office and others to retain postal address and other coded systems.

We believe there is a strong case for co-ordinating development of Land Registry, Valuation Office and local authority systems to provide the basis for a comprehensive national register of property information. Such a register could be organised on regional lines by the linking of computerised data sets.

OFFICE OF POPULATION CENSUSES & SURVEYS
The Census needs precisely located small areas for enumeration and for the output of statistics. There is also a need in the Census and in many other data sets for data referenced by postal addresses to be coded to zones for the output of statistics. For the most efficient handling of such data, a capability for automatic manipulation in the geographic dimension is required, for example, to produce statistics from different sources for compatible areas.

The postcode is a very convenient way of putting detailed area codes on any data that includes postal addresses and the postcodes are already extensively used through the Central Postcode Directory. But the existing postcode system has limitations for geographical referencing, so the Census Offices are proposing to improve the position by mapping the postcode areas and giving them geographical locations through National Grid references. In this way the postcodes, which can be reported by people in censuses and other returns, would be tied to the permanent geographical base of the National Grid which is precise, open to all and suitable for computer manipulation.

The Census Offices have experience of the postcodes system through the Central Postcode Directory and through use of postcode geography in the 1981 Census in Scotland, although the base in Scotland is not fully automated in the way now envisaged.

CENTRAL STATISTICAL OFFICE
I have consulted my colleagues in Departments (many of which will be replying to your request individually) and confirmed that for government statistics the use of postcodes is the preferred way forward in allocating data to small areas.

HM LAND REGISTRY
HMLR has always had a need for a means of locating individual properties on the large scale maps, in the majority of cases by the postal address. It has met this need by establishing and maintaining a street-card index, which provides a means of cross referencing postal addresses with public index maps or index map sections. If any other organisation were to produce a nationwide gazetteer of street names referenced to (say) national grid reference numbers, this would be of interest to HMLR. It is suggested that there is likely to be a demand for such a facility, and if so, OS should develop it as an integral part of its digital developement programme. In the meantime HMLR likely to develop its own facility, tailored to reflect the department's own particular database searching requirements.

METROPOLITAN POLICE
Numerous operational advantages could be derived by the police from linked/shared data sets. Many applications might prove to be of considerable value in the investigation of offences or for the prevention of crime. An example might be access to information about underground passageways to assist police in the investigation of a bank theft, where access was gained to a vault from underground, or to assist police in preventing terrorism by searching underground passageways for explosive devices. Similarly the provision by police of certain geographical crime data could provide substantial benefits, for example, to studies by Local Authorities and Central Government about the effect of environmental factors upon crime patterns which might influence building programmes.

The main constraints would be the confidentiality of certain crime information where disclosure might provide further criminal opportunities or adversely affect investigations.

ECONOMIC AND SOCIAL RESEARCH COUNCIL
We see considerable benefits from developing geographic information systems which bring together socio-economic data and environmental data. However, this will not always be, and may not usually need to be, linkage in the sense of integrated data: it may be sufficient for the respective types of data to be held in the same data store and merely cross-related. The techniques involved in linking spatial with non-spatial data are as yet at a very early stage of development and ESRC has earmarked special funds for expenditure by its Research Resources and Methods Committee on methodological research in this area. An early test will be the Rural Areas Database which is due to be established at the ESRC Data Archive. One obvious problem is the general lack of compatibility between the spatial units and referencing systems adopted by the various data collection agencies.

Major methodological problems in achieving greater compatability between the spatial units used by different bodies, of disaggregating data so as to approximate to common spatial units or using 'point' samples to relate to continuous spatial coverage of data (eg comparing 'point' sampled environmental data with 'continuous' data on population characteristics), interpolation of data for new spatial units and the 'ecological fallacy' of drawing inferences about individuals on the basis of aggregate data for areas.

NATURAL ENVIRONMENT RESEARCH COUNCIL

There are significant benefits to the environmental sciences from the linking and sharing of data sets. The consensus is that free and open interchange of data would benefit the scientific community with the added benefit that costs would also be reduced. Data sharing is likely to become of increasing importance in the future as data integration and analysis of multiple data sets grows. A current example of an ongoing NERC project where data sharing is an important factor is the BGS Moray Buchan Epidemiology Project where BGS, Aberdeen University, NuTIS and the EEC all benefit from access to pooled geochemical and geological data.

LOCAL GOVERNMENT BOUNDARY COMMISSION (SCOTLAND)

The 1971 Census of Population of Great Britain was spatially referenced using the national grid references for each dwelling: grid referencing was more subdued in the 1981 Census in England and Wales and in Scotland, post codes were substituted. It has been argued that the smallest post code unit consisting of some 12 householders provides sufficient spatial identification in urban areas but this is not so in rural areas with much more scattered populations. Post code units are much less flexible and indeed possibly less permanent than grid references which provide permanent spatial identification and are much more flexible, in that, using a digitizer, one could digitise the boundaries of an area proposed to be transferred from one electoral area to another and obtain the number of electors within the area. In order to develop this technique, electoral rolls would have to include 14 digit National Grid References for every address which would give the location to the nearest metre. One way forward would be first to provide National Grid References for each entry in the valuation rolls on a national basis: new buildings and sub-divisions of existing buildings would have to be recorded and added with new entries to the valuation roll. These grid references could be transformed to the electoral rolls. Central Government would require to institute such national grid referencing of data sets: the present provision of some data sets, fully or partially grid referenced and others identified by post codes is undesirable, and prevents linking of data sets on a spatial basis.

GRAMPIAN REGIONAL COUNCIL

The Council is also becoming increasingly aware of the requirement to co-ordinate data for individual properties and land parcels. Postal addresses while widely used are not amenable to automatic processing. In Scotland, each authority holds this type of data on the valuation roll and the basis already exists for a property data base using a system of unique property reference numbers. The opportunities are there but the basic problem exists in being able to devote the considerable resources required to establish such a system especially in smaller rural authorities with no major urban concentrations of population. Within its own service the Council is considering the introduction of a property referencing system which is unrelated to specific functional areas and which has some common link with spatial and attribute data. Information provided at low level can always be aggregated to suit local purposes; the opposite is often not true. There is no nationally agreed system of spatial referencing for the data the council acquires from other organisations. Only action sponsored by central government is likely to achieve this.

MIDDLESBOROUGH BOROUGH COUNCIL

The main problem in using spatially referenced information currently supplied by other organisations is the lack of a common geographic boundary for collecting data. For example census material is based on enumeration districts and ward boundaries which are not always consistent with estate boundaries and often divide a street or estate. The adoption of a common zone or spatial reference system for collecting all data, ideally at a small scale which can easily be aggregated could overcome some of these problems.

EDINBURGH CITY COUNCIL

The most effective way of overcoming these problems would be to aim for a universal system whereby all geographical information is expressed in terms of National Grid References. Such a system would have many advantages (i) most importantly, it offers great flexibility, since grid referenced data can be aggregated into an infinite number of larger areal units (eg wards) depending on what is required in any particular analysis. (ii) it offers precision and therefore lends itself to computer manipulation and analysis, which in turn may facilitate the identification of geographical patterns and relationships (iii) since the National Grid is a permanent and stable framework, the age-old problem of changing boundaries and inconsistent areal units would be circumvented, thus permitting a much more reliable investigation of time series (iv) in economic terms, the development of a unified National Grid coding system would avoid the need for duplication of effort by data collection agencies, which currently present information in a variety of areal formats for different end-users; this could result in substantially reduced costs.

HARINGEY BOROUGH COUNCIL

The London Borough of Haringey would look to Central Government to create a mechanism for the mapping and control of postcodes. To be of general use for the location of geographical information, this control should be separated from the day-to-day operation of postal services.

The London Borough of Haringey would therefore see as a major priority in the next 5 to 10 years methods introduced to ensure that each property is uniquely addressed and that the boundaries of the property are accurately defined by ownership. A major change such as this is clearly the responsibility of Central Government and probably the function of the Land Registry and Inland Revenue. It should be emphasised that unlike now, the information should be publicly available.

BOROUGH OF SOUTH TYNESIDE

Problems associated with use of spatially referenced information from other agencies generally arise from the referencing systems themselves, the inaccuracies in coding and the capacity for error in cross-referral to internally generated/used referencing systems (eg. post-coded data to Joint Information System's Unique property numbering/grid referencing). Unless rectification occurs in future planning, these difficulties are likely to increase (eg. if OPCS proposals for a geographic base for statistics relying on post codes are implemented for the next census without due regard to output and user access, the value of data will diminish). The problems could be overcome if users were consulted more closely in design processes of mapping and digitising unit boundaries. The Committee will appreciate, for example, the non-geographic nature of unit postcodes (ie. non-continuous strings of addresses which form convoluted shapes); their limitations in dealing with new, unmapped developments, and their potential for breaking down the current simple and extensively-used unit-built hierarchy of enumeration district/ward/Borough/County (Unit postcodes can cross Metropolitan District boundaries).

MERSEYSIDE COUNTY COUNCIL
We are confident that the development of a comprehensive address register that is capable of holding, maintaining and outputting information concerning the physical state of the borough, capable of application to all adress-labelled systems reducing duplication, and enabling access to strategic and management overviews is the correct way forward. We are basing all our initiatives for handling geographic information on this conceptual framework.

INTERNATIONAL COMPUTERS LTD
The linking of otherwise disparate data sets, such as those maintained by separate production systems, is one of the main 'raisons d'être' of geographic information systems. It allows more information to be derived from the data sets by using the spatial relationships of the the entities concerned. This accords naturally with human thought processes. In a sense the geographic information system must provide a 'window' on to largely existing data, without requiring data duplication.

INTERGRAPH LTD
Although the USA have been collecting data for these types of systems for a longer period than in the UK, they are not necessarily any further forward than the UK. In fact, if we could speed up the production of OS digital maps, there is a possibility that because of our physical area, one co-ordinate system, organized utility industry, we could, in a number of years, lead in a totally integrated geographically orientated information system.

TESCO STORES LTD
A major problem experienced in using spatially referenced data is the variation in the modes of referencing used. At Tesco the majority of information used is located by grid reference, however, data produced by local government or non-Census data produced by central government is often located by administrative units. This inevitably causes problems when several sets of data are being used jointly. One possible method of overcoming such problems would be through the use of digitised boundaries, whereby it would be possible to relate all types of areal units to grid references.

NATIONAL WESTMINSTER BANK
The key item on which the Bank will rely for direct marketing, analysis of its customer base and analysis of marketing opportunities generally *will be the postcode*. This provides access via commercially available conversion media to the *Acorn code,* which identifies market segments to which particular services should be offered. It also points to the grid reference, which could assume increasing importance as the means of establishing location of customers' homes and places of work, and relating these to the distribution of our branches.

Any proposed changes that might significantly alter the present postcodes should be considered with the utmost care. Full regard would need to be given to the effects on users' existing stores of information and to the cost of dealing with changes of the system.

COUNTY PLANNING OFFICERS' SOCIETY
The main problems of using information supplied by external organisations may be summarised as:

1. The lack of standard spatial reference units on each data record.

2. The impossibility of aggregating data into meaningful spatial areas, for example district council areas from job centre area.

3. Incompatible formats of computer held data leading to difficulties of data transfer.

COUNTY SURVEYORS SOCIETY
Great benefits are foreseen from the linking and/or sharing of data sets particularly with respect to public utilities, planning, and property data. Duplication would be reduced, staff time would be saved and transcription errors would be reduced to the minimum. Common standards of accuracy would need to be agreed, currency would have to be maintained with all information dated, and protection against inadvertent or unauthorised alteration of data would have to be ensured.

LOCAL AUTHORITIES ORDNANCE SURVEY COMMITTEE
The main benefits from linking and sharing data sets would be (a) it is the view of LAOSC that the National Grid should become the common geographic base for shared data sets; (b) saving of storage space in computer and manual files; (c) saving of staff time; (d) validity may be enhanced by reducing risk of using out of date or conflicting data; (e) software can be standardised. The main constraints are (a) liability for incorrent data, and the misinterpretation of data; (b) data must be on a central mainframe or inter departmental computer network; (c) privacy and confidentiality especially when personal files are linked; (d) common referencing standards; (e) common standards for data definition; and (f) software complexity and need for co-ordination. As an example the Committee definitely believe that the availability of data relating to underground utilities would be of great significance to local authorities in their property and planning data, which is the area where local authorities are most likely to start their use of spatial referencing.

It is the Committee's opinion that any geographic data used by local authorities in the future should be based on the National Grid, thus setting a standard by which all data can be referenced.

It is evident that in the past development has been sporadic and dependent upon the energies of individuals in specific local authorities. If this type of patchwork development continues it will prove uneconomic in terms of duplication, with all new users 're-inventing the wheel'. Therefore it is essential that standards and guidelines with Codes of Practice are developed, and more importantly, accepted by all potential users.

ROYAL INSTITUTION OF CHARTERED SURVEYORS
We would not wish to generalise, because each data-set needs to be considered separately, but provided all systems are tied to the OS base and the data is properly coded for accuracy, currency etc, we would see merit in systems which allowed central monitoring and co-ordination, but with regionally held data bases for easier and cheaper local access.

We have already discussed the need for standards but as this is central to the whole question of the organisation of data bases we would wish to raise it again here.

Ideally, the need could be argued for some central monitoring organisation which could check the quality of different datasets and see that they are linked as accurately as possible to the OS data-base. However, this seems an enormous and consequently impractical task. Therefore, the linking of data sets to the OS base must be left to each authority to do separately to the best of its ability.

ROYAL TOWN PLANNING INSTITUTE

Sharing data with public utilities would bring important benefits to planning. For example, the fact that a dwelling is occupied tells the local authority something about house building rates, progress towards housing targets and the distribution of demand for land. At the moment, most local authorities gain this data from either the rates system or building control inspectors. However, if rates are abolished as a means of local taxation, and an alternative property tax not introduced, the only remaining source of such data would be the electricity boards. In any event, electricity boards are likely to be a more accurate source of such data since all dwellings are supplied with electric power which must be paid for from the day of installation by the relevant board.

The main constraints to sharing data are:

(1) Lack of consistent and compatible quality and collection of geographic reference.

(2) The need to ensure easy access to the information whilst not impinging on the provisions of data protection legislation.

(3) Difficulty in gaining agreement from potential data sources to regular supplies.

(4) Technical constraints, in terms of incompatible equipment or software or data structures.

(5) Costs and resources.

Sharing data is constrained by the lack of standards. If different types of geographic information are to be merged or exchanged within computerised systems then it is vital that they should be compatible with one another in a number of important respects. These include the way data is organised, the quality of that data, the terminology used, and the way in which the information is spatially referenced. It is therefore the Institute's view that there is a clear need for a Code of Practice to cover the important subject of data exchange.

HOUSE BUILDERS FEDERATION

In addition the Federation identifies a clear need for a common level of spatial compatibility of the wide range of data. It believes that the planning and provision of services and land uses would be greatly enhanced: Such provision is inevitably based on judgement about levels of requirement and the Federation considers that greater geographical compatibility would lead to much better informed policy making. HBF believes this could be facilitated through the establishment of a common building block of data (or a Basic Spatial Unit) for geographic and other related information. This could in turn be used to develop a hierarchy of related information systems. The establishment of such a framework would allow the wide range of official Government published data to be reconciled and ultimately could lead to the integration of these official data with the wide range of private sector data sources that appear to be available and which can usefully suplement official Government statistics. A sufficiently small 'building block' would clearly facilitate the aggregation of data in a flexible manner to meet a variety of policy making needs in central, and local government and commercial areas.

An additional problem is created by the fact that in many plans policy areas are used to develop specific aspects of general policies with reference to fairly localised areas: These policy areas may often cut across district council boundaries. It is therefore often particularly difficult for HBF and indeed other organisations and individuals to make an effective analysis of local authorities proposals, since published data is not generally available at the particular level of spatial resolution of the policy area. Indeed this begs the question how the local authority have derived those policies in the first place!

ROYAL SOCIETY

The principal obstacles to linking different datasets are the lack of common standards and the lack of a comprehensive digital database of OS maps; others include no or poor documentation on the characteristics of different datasets, with particular reference to their reliability and to the potential accuracy of the data they contain; the fact that many data, especially those collected in the past, are not in digital form (a major concern for geologists); the inaccessibility of data because of considerations of confidentiality or official policy; the use of different sampling frames; and the sheer cost of acquiring such data. Many of these difficulties arise whether data are in analogue or digital form; but it is impracticable to compare or relate most datasets for any large area unless they are in digital form.

ROYAL GEOGRAPHICAL SOCIETY

Geography is greatly concerned with linking data of different kinds or relating data for different periods in the examination of change. There should be no major problems in linking or sharing data provided that there is the essential common map framework (by the Ordnance Survey in Great Britain) and that common standards for all geographic information are accepted. Other major constraints to the linking or sharing of geographic information data in Great Britain are: the disparity of common standards among agencies collecting data (in contrast to the situation in other nations such as the United States, Germany, Canada and Australia which have taken the initiative nationally to have geographic information collected and held to a common standard); ignorance of what other organisations or agencies hold and the characteristics of their data; lack of expertise particularly in handling large data sets; cost of data; fear of new technology; possessiveness towards the data collected and an unwillingness to consider the needs of others.

BIRKBECK COLLEGE, UNIVERSITY OF LONDON

Many significant and time-consuming problems can occur when using information supplied by other organisations. The *lack of consistent data formats* and transfer formats are notable, as are the variations *in accuracy* of data. Another problem in using data supplied by outside bodies is that the base unit used for *spatial referencing* often differs from one data set to another, and this can restrict and complicate data integration. Perhaps most frustrating of all is the common lack of clear and complete documentation, itself produced to well defined common standards.

UNIVERSITY OF NEWCASTLE, CENTRE FOR URBAN AND REGIONAL DEVELOPMENT STUDIES

The principal outcome from this review of spatial referencing in official data series must be a strong endorsement of the present moves towards postcoding and the development of flexible aggregation systems based upon it. However such a narrow technical advance must be part of a very much broader initiative if the full benefits are to accrue from the expenditure involved. In particular, other changes needed are:

- control of the postcode system to be taken by an agency responsible for national spatial referencing

- all postcode units to be territorially defined, digitised and any changes documented

- all post-1970 administrative and statistical areas to be defined as aggregates of postcodes

- all official datasets to be computer coded at the finest possible level of areal resolution

- all datasets to be re-evaluated with regard to the minimum levels for which data can be published (including user-specified aggregations)

- new Official Standard areas to be promulgated, at various levels to encourage the cross-fertilisation of data sets

- new administrative and statistical areas to be aggregations of Official Standard areas

- new legislation enacted, if necessary, to facilities the release of anonymised samples of individual Census and industrial establishment data

- new initiative launched to reduce inter-Departmental inconsistencies in spatial data referencing systems and coding practices

UNIVERSITY OF BRISTOL, SCHOOL FOR ADVANCED URBAN STUDIES

A major problem for researchers and practitioners seeking to integrate several data sets on a spatial basis is the sheer variety of the basic spatial units upon which data is collected and/or made available. This is especially a problem at the small area level. Different sorts of data of necessity relate to different kinds of spatial unit. For example, thematic data tend to be aggregated on the basis of nesting hierarchies of administrative units (eg. LA's, wards, ED's); specific events of phenomena by specific geographic position (eg. postal address or location on a topographical map); flows assigned to networks (eg. roads, utility networks). The picture is further complicated in that within these categories the choice of unit will depend on the immediate interest of those collecting a particular set of data as well as confidentiality constraints so that there is no commonality of spatial unit even for the same type of data (eg. employment data areas differ from population census areas). A major improvement in the utility of data sources would be achieved by the adoption of a common framework of geographic referencing.

However it is not realistic to suggest the adoption of *common spatial units*, not only because of the different purposes and interests of data collectors and providers, but also because secondary user themselves require the integration of data on a variety of spatial bases, depending on the focus of their analyses. For example, demographic data may be needed for travel zones to link with travel flow information, for school catchment areas to link with data on existing school capacities, for water authority districts to link with network loads and capacities. These requirements vary both over time and according to the level of analysis. Thus what is needed is a flexible location referencing system which will facilitate data transformation from one spatial unit to another and permit the *linkage* of one type of spatial data to another. However there are still inter-related questions concerning (i) what kind of location referencing will best meet these requirements; and (ii) the optimum level of resolution to meet user needs.

KINGSTON POLYTECHNIC

There exists a significant data integration problem which is preventing the widespread use of spatial data in many disciplines. This arises primarily from the sortage of data in *incompatible formats,* from the *restriction on access* to certain datasets, and from problems associated with *linking* together widely distributed spatial databases. The requirement to combine traditional vector coded cartographic information with raster based high resolution imagery poses other unresolved problems. For example, the storage of cartographic data in a raster format leads to a data explosion, while the extraction of map information in vector form from imagery can not yet be performed reliably despite much research work.

There also appears to be a major user interface problem associated with spatial information systems. This problem has two components: one is the difficulty experienced by a data user in selecting the most appropriate datasets for a given application. The other is the problem of presenting spatial data to the end-user in a way which conveys the maximum visual information so that even non-experts can make effective use of the data on display. This is essentially a problem of automating graphic design, and of resolving graphical conflicts arising from combinations of different datasets with overlapping features.

US DEPARTMENT OF THE INTERIOR, GEOLOGICAL SURVEY

The obvious benefits of sharing data are to eliminate the need (and expense) of collecting information over again and to expand the horizons of an analyst to use information that might otherwise been excluded. A minor constraint is the variation in formats among various data bases, however this is not insurmountable. The major constraint concerns data quality. What level of confidence can be placed in the data obtained from another source. If the data are the results of an interpretation (eg. photo interpretation of land use), what level of confidence can be placed on the interpretation. Most data sets being distributed today contain no reference or description of its legacy or quality. Indeed, there is no consensus among scientists working in this field as to what type of data meaningfully describe data quality. Until this is overcome, the 'lack of faith' in anothers data will be the major constraint in sharing data.

E.   Awareness

ORDNANCE SURVEY

One possible way forward would be for the co-ordinating agency to arrange for the development of demonstrator projects (one based on large scale data and one based on medium or small scales data) in order to stimulate widely based user interest. These projects might attract some level of central government support on the grounds that GIS have the potential to produce benefits at the national level through better planning and faster more accurate decision making.

NATIONAL REMOTE SENSING CENTRE

We would therefore recommend that a demonstration project be set up with the following aims and using existing technology:

Educate potential users in the applications of remotely-sensed data within the existing spatial information technology and allow individual managers to estimate the value of geographical information systems to their organisations in terms of day-to-day activities and strategic planning needs. Unless some plans are made to provide an educated and informed user community there seems little prospect of serious use being made of whatever spatial data become available.

Provide a means whereby management could assess their requirements in terms of capital investment, expenditure on training, and staffing levels.

Provide an opportunity for problem areas to be identified and priorities to be assessed so that R& D might take place against a more realistic background.

Allow organisations such as the Ordnance Survey, the Meteorological Office, the Soil Survey, the National Remote Sensing Centre and all other agencies concerned with the collection of environmental data to collaborate in the production of a body of digital spatial information that would form the basis of a successful demonstration project. Liaison between these organisations and potential users would be essential. The data so generated would be invaluable in the testing and comparison of algorithms. This approach would permit a modest start to be made; later efforts could be devoted to filling in the gaps.

SOUTH OXFORDSHIRE DISTRICT COUNCIL
There is a dismal lack of awareness nationally with regard to the huge potential of computer mapping and corporate land and property systems. Very little has been done and this situation must be rectified quickly and urgently.

CHESHIRE COUNTY COUNCIL
There is also the problem of a lack of awareness amongst many of those who collect and use data that there is the potential for geographically manipulating the data in ways that will help in planning service provision and service delivery.

Given appropriate mechanisms for managing, transmitting and analysing geographic data easily and cheaply, the use of such data by Country Council departments for planning and operational purposes seems likely to grow. A number of future developments have already been identified in Cheshire.

INTERNATIONAL COMPUTERS LTD
How can we achieve a greater awareness of the true potential of geographic data and geographic information systems? It is suggested that the way forward is through some central agency having the remit to institute:
Awareness programmes; Modification to relevant professional training and education; Demonstrator projects; Creation of accessible data services that would be chargeable to users; Publicity for innovation (whether UK based or overseas).

CACI LTD
The major area for the future must be an integrated approach for data handling, analysis, graphics and interpretation. However, in our opinion, the level of understanding in commercial organisations of the power of such systems is limited and consultancy help is paramount. This should extend into professional and academic training, particularly in respect of marketing and property areas.

KINGSTON POLYTECHNIC
The lack of exploitation of spatial data calls for several courses of action in our opinion. First of all, there is a need to make the potential user community *aware of the benefits* of spatial data and geographic information systems. This particularly applies to commercial companies involved in transportation, utilities, surveying activities, civil engineering, estate management etc. To some extent the recent research initiatives such as the large-scale Alvey demonstrator project on mobile information systems may take some impact in a limited area. Secondly, there is a need to make *available digitised Ordnance Survey* maps of all major towns a at scale appropriate to local commerce ie. 1:25000 or better. Ideally digitised maps should be made available together with appropriate software packages for very low cost. Widespread use of spatial data will only come about if datasets and spatial database management software are readily available.

UNIVERSITY OF MANCHESTER
The development of very high capacity storage media is having considerable impact on the future of Geographical Information Systems. The BBC Domesday project will do much to popularise, and to alert the users of geographical data, to the possibilities of media such as Video disks and CD ROMs. Each of these offer the opportunity of publishing very substantial databases which can then be used by a large variety of end-users. The availability of large databases, and the software to interrogate them, on a cheap and accessible medium will go a long way towards obviating the need for on-line access to centrally held databases.

US DEPARTMENT OF THE INTERIOR, GEOLOGICAL SURVEY
Initial resistance to GIS technology stems from both a technical viewpoint as well as an economic one. On the technical side a user must be shown not only that the technology works, but also that it works on his particular problem. Economically, the user must justify the relatively high startup costs associated with implementing GIS technology.

Well documented GIS applications demostrations will show the naive user that the technology does work. They may not be enough, however, to convince a user that GIS work on his particular application. In such cases, a cooperative demonstration project between a lead GIS agency and the naive user may be necessary. In our experience, such cooperative demonstration projects can be quite effective in making users aware of the potential benefits of GIS technology.

F   Education and training

ORDNANCE SURVEY
Clearly a computer literate staff will have some impact on future developments but computer literacy alone is not the key to success. Spatial awareness and an understanding of geographic information are equally if not more important. Experience has shown that it is more effective to teach a professional or technical specialist how to design and write software packages that it is to convey the nuances and particular problems of specialisms to pure computer scientists. Education is the key to success.

The education of designers is very much a chicken and egg situation − without existing GIS it is difficult to illustrate the practicalities of design, but without designers it is difficult to create suitable GIS for training purposes. Education of designers is therefore still an evolutionary process. There is potential eventually for the establishment of Masters level courses at universities to encompass topics such as spatial data referencing, the handling of large data sets, raster to vector conversion, data communications, computer hardware and software developments, and geographic information in general.

The training of managers and users is more likely to be on the job or through short courses offered by systems builders. These will have to be more than just instruction by rote, they will have to consider in some depth the implication of combining various datasets and the costs of generating specific outputs.

HM LAND REGISTRY

From HMLR's own experience, it is clear that there is at present a shortage of personnel with the necessary background and experience required to design and implement a computerised geographical information system. Presuming that the requirement for such systems will grow, it appears that action is needed to promote the establishment of relevant educational and vocational training programmes and related facilities, and to generally encourage computer literacy. Educational institutions, industry and the appropriate professional organisations all have a part to play in meeting these needs.

MINISTRY OF AGRICULTURE, FISHERIES AND FOOD

Any substantial development in the GIS field in the future will create a requirement for specific in-depth training to develop and operate the systems. The ADAS Data Flow Study identified the importance of establishing efficient and appropriate data flow structures, and skills in this form of systems analysis will need to be applied to spatial as well as non-spatial data.

THE ELECTRICITY COUNCIL

At present training needs are being satisfied in general terms by suppliers of computer aided draughting equipment. However, the scope of this training is limited and inhouse customising of training is common. As system complexity increases training packages will need to be designed to cover a wider spectrum of needs. These would range from general awareness programmes, for staff whose work may in some way be affected in the future, to courses for specialists aimed at providing complete and impartial information upon which technical evaluations of systems and equipment can be based.

Educational institutions could play an important part in designing the training courses together with the training establishments of the organisations concerned. This would avoid the duplication of effort at each user centre and with the frequency of new releases of software, updating and refresher courses could be accommodated.

SOUTH OF SCOTLAND ELECTRICITY BOARD

Educational and training needs are largely being met by vendor systems suppliers and in national and international seminars. However, this in many respects is deficient in that each system user has largely been required to train in-house the majority of their operators, also vendor training covers the system as a whole and may not cover the sphere of operation of the user. In other words each user has to customise the systems to meet his own particular needs and consequently develop his own training courses.

Educational establishments, e.g. central institutions, should be funded to set up training courses to meet individual user needs. Particular systems should not be duplicated at different institutions in any regional area. Refresher courses could be organised for new software issues.

SYSSCAN UK LTD

As an employer of staff specialising in digital mapping and geographical information systems, SysScan is disappointed at the lack of awareness by many recent graduates of the scope of this technology. It is regrettable that we can readily find cartographers and geographers; that we can readily find computer scientists, but that graduate staff with an awareness of both disciplines are very rare.

INTERNATIONAL COMPUTERS LTD

If the full benefits of using the geographical context of data are to be realised, a combination of training as part of professional qualifications, relevant short courses by Universities, seminars by organisations such as BURISA and the British Cartographic Society must be assembled, co-ordinated and marketed.

INTERGRAPH LTD

In the present climate of fast changes in hardware and communications, it is very difficult for educational bodies to keep totally (practically) abreast of the current situation as the advent of a new CPU, another level of ISO, optical disk etc., can tip the financial considerations in a variety of ways for a commercial company.

Obviously the younger generation is more computer literate and will, therefore, grasp the potential of geographic information systems more easily. But it will still require the experience of the professional person such as the gas engineer, the planning officer etc., to know how to use the information and to analyse it to do his job more effectively. In the initial stages of data collection, it is even more important that practical, experienced engineers are consulted about database design and future uses.

BKS LTD

It is essential that our education system reviews its procedures to ensure that its students can relate to current industrial trends and are aware of industry's problems. This could be achieved by promotion through industrial liaison units for students to take on industrial projects as part of their curriculum and a pre-requisite to their qualification assessment. This is especially significant as state of the art system costs generally exceed university budgets and familiarisation with existing techniques is only available through industrial institutions.

This could lead to the development of co-operative projects harnessing the expertise of the university with industry to complete commercial contracts and further the acceptance of university developed techniques.

Major educational and training requirements arises from the needs of the end user. It is a fact that the end user at this time is not well informed on the subject of geographical databases. If we are to establish a national database system the end user has to be convinced of the viability of such a system and he will only be convinced by education.

Such education and training needs to be provided on a national scale by government utilising the experience of database system suppliers and companies experienced in the field.

LOCAL AUTHORITIES ORDNANCE SURVEY COMMITTEE

Local authorities provide training for their staff in many aspects of their work, and most authorities have a commitment to staff training. Experience is limited in the use of geographic data, especially amongst management, and a shortage of officers in local government with the necessary expertise to take the lead in the introduction of a Geographic Information System. There is, however, a very great interest in the potential advantage of such systems for local government and other public authorities. There is a need for a corporate as well as a departmental approach, although the departmental benefits are easier to identify. There is also a need for education and training at all levels, and it may be relevant to suggest to those organisations which concentrate on management education that the corporate benefits of a Geographic Information System should be included.

COUNTY PLANNING OFFICERS SOCIETY

The whole issue of training for computer technology is currently being examined by the County Planning Officers' Society's Use of Computers Working Group by questionnaire survey of its membership. Initial findings show that universities and polytechnic's in association with local authorities could, and in several cases already have, collaborated in providing training opportunities. Interestingly, whereas 2 or 3 years ago general computer awareness training was the norm the present and likely future, trend is towards 'self help' using tutorial packages, guided where necessary by experienced users. The implications are the need to make computer equipment available for training purposes and to allocate time for staff to develop at their own pace.

Complementing this 'self help' is the need for specific courses based on using specific packages plus more evaluative workshops which take stock of the value of the available technology. The recent British Urban and Regional Information Systems Association (BURISA)/Birbeck College workshop on 'Computer Mapping for Local Authorities' is an excellent example of what can be achieved.

ROYAL GEOGRAPHICAL SOCIETY

Although a welcome start has been made in the development of advanced training in these fields, for example several MSc courses at British universities, there is a lack of appropriate hardware and the software needed as graduates apply these skills. There is also the short term problem of the training in these new skills of the older generation who are often in a position to encourage the exploitation of these systems. This can be remedied by short courses but it will need example and encouragement from Government. A successful example has been the National Science Foundation in the United States funding of short courses in the 1960s for senior academic staff unfamiliar with the developments in computers and the new skills required.

ROYAL SOCIETY OF EDINBURGH

The Society agrees that there is a shortage of those with appropriate skills and believes that it will be necessary for Government to provide ear-marked funds for postgraduate studentships for taught courses in digital mapping, remote sensing and geographical information systems, such as those now running at Glasgow, Edinburgh and Dundee Universities. Scotland also offers good opportunities for two-way exchanges between those working in these fields in academic, commercial and government organisations; but such exchanges will require encouragement and pump-priming to overcome initial inertia.

Another major issue of concern, not apparently considered in the issues noted by the Committee, relates to the statistical reliability of geographic data. There appears to be widespread ignorance among users of spatially-distributed data of the recent upsurge of interest in this area by statistical theorists who have shown the inappropriateness of existing methods that are widely used. This deficiency is also apparent in education and training in the handling of geographically-referenced data. As well as the provision of training in computing, especially in the handling of large data sets and in the use of commercially available software, there is also a need for training in statistics and a need to harness the expertise of statisticians in research and development in this field, especially in relation to the development of geographic information systems and the handling of remotely-sensed data.

REMOTE SENSING SOCIETY

The introduction of GIS will require investment in the education and training of potential users. It will not be sensible to rely on the diffusion of information from new recruits. Education and training may be particularly important where new types of data are being incorporated with other, more familiar, data. For example, the user may be familiar with conventional map data but may lack familiarity with classified remotely-sensed imagery. Education and training carried out by universities and polytechnics should not be expected to proceed without clear directives regarding level and content of courses. Such direction may come from a central organization with responsibility for education and training. Piecemeal or ad-hoc arrangements will not suffice. It is difficult to see how a Government ministry could effectively control teaching in universities and polytechnics, which are largely independent organisations. A voluntary professional society or similar body covering the broad, interdisciplinary field of spatial information technology might be able to produce an agreed or model syllabus to guide teachers in further and higher education. Such an organisation might additionally provide a useful forum to bring together specialists who at the moment are scattered over many academic departments and research units, both private and public sector. The range of specialisms is considerable, and includes computer science, psychology, electrical and electronic engineering, mathematics, statistics, geodesy and cartography as well as remote sensing.

ROYAL SOCIETY

Education and training in matters relating to the acquisition, processing and display of spatially referenced data take place in most universities and polytechnics — indeed, all courses in geography are by definition concerned with such matters although the extent to which such training is computer based varied widely. In remote sensing and digital mapping the field is less wide and relatively few specific courses in such subjects are offered. Employers have reported difficulties in recruiting suitably qualified staff to work within the general fields of computerized spatial data management. Many are also unable to identify training courses appropriate to their needs. Undergraduates studying in fields associated with the handling of spatial data need to be made more aware of the problems and of the systems that are currently available.

A great variety of applications exists, ranging from large-scale local surveys to the processing and analysis of satellite imagery, but the emphasis has generally been on solving identified problems rather than on developing an educational strategy. Recently, postgraduate courses in GIS, remote sensing and in digital mapping have begun to appear, but they are still in their infancy. Any general evaluation is therefore premature.

The relation between established courses and the needs of Government and of industry follows no coherent pattern. In many instances, the academic community, in both its teaching and its research, is some years behind the more significant developments that have taken place in the commercial and government sectors. There is little coordination of research activities between educational institutions and the ambiguity of responsibility between NERC and SERC, to which reference has already been made, creates difficulties. The Royal Society's evidence to the House of Lords called for the creation of a Joint Research Councils Committee to promote and facilitate research in remote sensing and digital mapping, and this remit needs to be extended to the handling of geographic information generally and to the development of GIS specifically.

Courses designed to assist those in employment to understand the new technologies suffer from a lack of equipment and funding, a lack of training and experienced academic staff and from highly specific requirements of employers with relatively small numbers of trainees. Members of the academic community need to work in or with industry and to experience at first hand the nature of the tasks and procedures that are being adopted by organizations involved in GIS and in the handling of spatially referenced data. Here, as in many areas, exchange of staff on short-term or long-term secondments between academia, industry and government can only be beneficial. Until there is greater standardization of both hardware and software, no institution will be able to cope on its own with this diversity. Centres of excellence need to be established where the more sophisticated systems can be installed. Closer cooperation is needed between academic establishments; for instance, within London, where the range of facilities and expertise that exists among institutions far exceeds that which any one of them alone can offer, but where there is already established an M.SC course in remote sensing that has GIS as one of its options. To exploit such cooperation the confidentiality and copyright on software and data will need to be relaxed so that both can be made more transportable.

BIRKBECK COLLEGE, UNIVERSITY OF LONDON

Considering the importance of geographic information and the great lack of specialists with the ability to understand and handle this data, it is imperative that more higher educational courses be run and that money is invested in training such people. It is still rare to find employees with a comprehensive background in geography who are also computer literate. Our preferred approach is to mount short courses.

UNIVERSITY OF DURHAM

Production of graduates with experience in high technology spatial data systems is well below demand. More image processing work stations are needed. There is too little support for postgraduate work on GIS.

UNIVERSITY OF MANCHESTER

We are not aware of any full computer cartography courses in higher education. There has been an increasing awareness of the need to offer training in these fields which has been reflected in the content of general courses in many geography departments. The need is to develop a limited number of more specialised centres offering courses of a multidisciplinary nature. Such centres should have a regional spread so as to maximise the potential connection (both for advice, research and continuing professional development teaching) with practitioners in local government and other regional public bodies such as health authorities. Geography departments could offer a valuable umbrella for such multidisciplinary work which should include computer scientists, social administration, geographers, community medicine, business studies and others. The emphasis of training should be towards management applications of spatial information systems.

LUTON COLLEGE OF HIGHER EDUCATION

As the number of applications of geographic data increases and the amount of data which is held in geographic form increases, it is our view that the capture, processing and presentation of the information will be carried out by technicians. These technicians will require both sound basic training and sufficient flexibility to adapt to rapidly changing techniques of data handling.

Whilst students entering technician and higher technician training courses may have basic computing graphics their ability to handle spatially referenced data is likely to be limited. The current application of computers in geography teaching is primarily in modelling and data processing of a statistical nature. Remote sensing and satellite imagery interpretation are beginning to appear in school syllabuses but strengthening this position will require investments of time for teacher training and of money to acquire materials and equipment at a time of scarce resources. The linking of computer and satellite data which may be achieved at present in a few schools is a long way off for the majority. Consequentially we shall expect to have to provide training commencing from the basics over the next five to ten years at least.

NORTH EAST LONDON POLYTECHNIC

Most courses dealing with geographic information will need to expand their treatment of databases and make much more use of them as they become more generally available. Some institutions will need to be encouraged to develop more specialist courses at postgraduate level to ensure that sufficient students with the correct academic background are available for government institutions, commerce and industry. Mid-career courses will also play an important role in providing this background. Courses of this nature inevitably need high calibre staff and equipment comparable to that used in industry. Much greater cooperation between industry and educational establishments will be needed to ensure that these two conditions are met. Government policy should be such that it positively promotes this requirement.

G Research and development

DEPARTMENT OF THE ENVIRONMENT
The Department's experience to date suggests that the future main requirement in geographic information systems is likely to be for more efficient database management tools. Simple menu driven programs will be needed to encourage wider use of geographic information systems.

MINISTRY OF DEFENCE, MILITARY SURVEY
Development in the acquisition of remotely sensed data, and its analysis, together with advances in ADP techniques and communications systems, will all have a significant effect on Military Survey's production techniques in the future. Areas of particular importance to increased efficiency are:

Databases. Military Survey is examining the possibility of generating a product independent geographic database from which a range of geographic products can be derived and maintained. This is termed the multi-product operation (MPO) concept. Developments we are now seeing in data structure and in relational database techniques will make MPO a practical concept.

Raster Technology The use of raster scanning will provide for rapid capture of data to populate the database as well as providing for rapid reproduction of existing maps. Developments in raster plotters are also likely to lead to an improved output from such databases.

Expert Systems. The development of artificial intelligence and knowledge base systems (AI/KB) for change detection, pattern recognition and feature extraction will more effectively utilise the new digital sources of remotely sensed data such as SPOT and LANDSAT. The introduction of such expert systems should provide considerable assistance in the maintenance review procedures of products and will also assist in automating the data capture process.

MIS. The increased use of management information systems within Military Survey will lead to greater efficiency in the planning and control of production resources.

Communications. The developments in local area networks and in narrow and broad band communications will allow for greater integration and control of production systems, rapid access to databases and real time data transmission.

NATIONAL REMOTE SENSING CENTRE
The main project recommended is the conceptual design of an IGIS using knowledge-based techniques, novel spatial data structures such as hierarchial data structures, and appropriate associated hardware. We feel that it is essential to base the system design upon a substantial UK user needs study. The latter could draw heavily upon the user survey being carried out by the DoE Committee of Enquiry and on the user surveys of Thomas and Stewart 1985, EARSel WG13 Report 1985, Green et al 1985. However, in order to ensure that the needs extracted are sensible, it is also proposed that the study contain an educative element aimed at educating potential users as to the benefits of a GIS, prior to eliciting their needs. Longer-term GIS education is also required, perhaps in the form of a GIS user 'club'.

METROPOLITAN POLICE
Future R&D work should concentrate upon the identification and adoption of a common approach to information handling systems so that as much information as possible is available to all who may have a legitimate interest in it. Funding should be sought from all bodies and organisations who stand to benefit from this work.

NATURAL ENVIRONMENT RESEARCH COUNCIL
Technological developments will make major contributions in a number of areas, including:

(i)   The development of novel forms of data structure suitable for holding all forms of spatial data in a compact form allowing efficient retrieval and processing (eg hierarchical data structures such as quadtrees).

(ii) The use of knowledge-based techniques within various subsystems of the IGIS. These might be incorporated into the data capture stage (eg knowledge-based image segmentation to improve classification accuracy), or the data retrieval stage (to enable fast database search), or might provide a friendly man-machine interface (eg to enable a user scientist to use an unfamiliar system in the most effective way).

(iii) Improvements in automatic cartographic data-capture by raster scanning and line following techniques. This is an important point as improvements in both the techniques and the cost of automatically digitising existing maps would avoid potential bottlenecks in the generation of sufficient digital vector data for the IGIS.

(iv) Improvements in basic machine architectures, to allow much higher processing speeds than are currently achievable from typical general purpose serial processing minicomputers. These improvements will probably involve using Very Large Scale Integrated Circuitry in non-serial architectures (eg multiple-instruction, multiple-data-stream systems).

(v)   Enhanced methods of on-line and archival storage for the massive data influx, including high-density magnetic storage as well as totally new media (eg optical discs).

(vi) Efficient network facilities at a national and local level. User requirements will range from a quick-look capability to the dissemination of very large data sets, and will dictate transfer rates according to task.

(vii) Remotely sensed data is still largely experimental. To date insufficient emphasis has been placed on the development of measurement methods and image analysis techniques aimed at achieving an operational applications capability. Considerable additional research is needed on the development of applications before the IGIS raster chain can be considered to be operational.

ESRC DATA ARCHIVE
As far as future prospects of gaining access to novel and useful data types are concerned 2 distinct areas stand out as holding the greatest promise for advances in information system technology. The first is the use of remotely sensed imagery for mapping land use and environmental data. The second is the development of more sophisticated interpolation techniques designed for estimation of missing data or data which are available only in aggregate or sampled formats.

BRITISH TELECOM
Technological advances which might materially increase efficiency and reduce the costs associated with implementation of computers graphics includes developments such as:

- new mass storage devices to replace magnetic disks and tapes (magnetic tapes demand too high a level of system management for on-line general applications);

- fast data capture, subjecting existing records to scanning processes;

- low cost work stations with a 'stand alone' capability;

- a 'read only' facility for use by mobile staff. Also, for the exchange of information in the field with other Utilities.

INTERNATIONAL COMPUTERS LTD
Technological change can be confidently predicted to continue and result in constant revision of cost/benefit ratios in this application area as in other. Areas that offer significant potential include:

Non-serial processing architecture

Storage

Workstations

Networks

Logic programming

ROYAL SOCIETY
Advances are needed in several areas of computer technology, including data entry, systems interfacing, data integration and output, and knowledge-based systems. The computer-systems industry should actively pursue those advances. Concerning basic research, proposals may fail to be funded by research councils because they are not central to a particular research council's interests. The research council structure is quite unsuitable for ensuring that such research of national importance is undertaken. Earmarked funding is needed, in line with the recognition by the Advisory Board for the Research Councils of the importance of techniques for handling spatial data. As called for in the Society's submission to the House of Lords, a Joint Research Councils Committee is needed to promote and facilitate research in remote sensing and digital cartography, and this should include:

(a) the use of radically new data structures, such as 'quad trees', to improve the efficiency of both storage and processing;

(b) the use of knowledge-based systems in the capture, analysis and retrieval of data;

(c) improvements in data capture by raster-scanning and automatic line-following techniques, and reduction in the costs of these techniques themselves;

(d) greater 'user-friendliness' in systems, so that packages can be run readily and databases accessed with a minimum of skill;

(e) improvements in basic machine architecture, particularly with the implementation of parallel processing;

(f) improved performance (reliability, speed, protocols) of public data networks and improved compatibility between machines — these will be essential if users are to have effective access to national geographic databases, whether these are dispersed or held centrally;

(g) enhanced methods of cheap storage of data in large volumes — these will include high-density magnetic storage as well as new media (eg. optical discs); video technology may play a role in the distribution and display of cartographic data although such analogue images will be of less importance for the management of integrated GIS.

Another major area of concern, not apparently considered in the issues noted by the Committee, relates to the statistical reliability of geographic data. There appears to be widespeard ignorance among users of spatially-distributed data of the recent upsurge of interest in this area by statistical theorists who have shown the inappropriateness of existing methods that are widely used. This deficiency is also apparent in education and training in the handling of geographically-referenced data. As well as the provision of training in computing, especially in the handling of large data sets and in the use of commercially available software, there is also a need for training in statistics and a need to harness the expertise of statisticians in research and development in this field, expecially in relation to the development of geographic information systems and the handling of remotely-sensed data.

It is government policy that funding for basic research should come from the research councils. The Natural Envirnoment Research Council, the Science and Engineering Research Council and the Economic and Social Research Council all currently fund research of relevance to spatial data handling. However, experience by scientists suggests that too often research proposals in topics such as spatial data handling that are not central to a research council's interests or to those of its committees often get passed from one Council to another owing to ambiguity over responsibility for funding. Moreover, they tend to be unfavourably regarded by the specialist committees when they finally come to be considered and so fail to be funded. The research council structure is quite unsuitable for ensuring that such research of national importance in undertaken. The Advisory Board for the Research Councils recognized the importance of techniques for handling spatial data in its 1984/5 submission to the Secretary of State for Education and Science, but although this report was generally accepted by the Secretary of State, additional resources have not been made available. We regret this decision; without such earmarked funding, the research is unlikely to be undertaken.

ROYAL INSTITUTION OF CHARTERED SURVEYORS
There is a wide range of possible R&D projects which could be undertaken, many of which have been mentioned earlier in our evidence. These include:

• Further research into user needs for spatial data

• Formulation of common hardware and software standards

• Development of data transfer standards

• Development of nationally agreed spatial referencing systems

• Research into database theory applied to spatial data

• Development of faster digitising systems — such as raster scanning and also faster/more accurate raster plotters.

• Faster data transfer systems via optical discs, fibre optics etc.

• Development of standards for symbols and presentation

We believe that government funding should be made available for selected areas of research. This being channelled through the proposed independent Body or on their recommendations. Some of the funding may be made available through existing research councils, but we see a need for a Joint Research Council, as the present split between the Science and Engineering Research Council and Natural Environment Research Council is unsatisfactory.

ROYAL STATISTICAL SOCIETY
The collection, summarizing and analysis, including interpretation, of any quantitative information are statistical procedures and require statistical skills if the eventual outcome is to be accurate, relevant and efficiently achieved. Even where data are to be collected, recorded and analyzed on an entirely automatic basis, statistical consideration should form an essential part in constructing a proper system. However, as regards spatially referenced data, the advice of a statistician is sought all too infrequently. We recommend that research council awards should be conditional on the involvement of a statistician.

REMOTE SENSING SOCIETY
We suggest that GIS research should be taken out of the present Research Council funding scheme simply because it is multidisciplinary and involves collaboration between scientists working in many different areas of knowledge (including engineering, physics, computer science, mathematics, statistics, and the user disciplines such as geography, geology, hydrology, argicultural science, forestry, botany and pedology). The structure of the Research Councils seems to make collaboration between research workers in these areas more rather than less difficult.

UNIVERSITY OF BRISTOL, SCHOOL FOR ADVANCED URBAN STUDIES
The kind of project that we have in mind is illustrated by the GIS that is currently under discussion between this department and the Bristol/Avon health authorities. Such as GIS would seek to establish geographical controls in environmental health problems in the Bristol/Avon region through the integration of individual health records, aggregate socio-economic information for areal units and environmental parameters extracted from remotely sensed data. The project will demonstrate and evaluate the rewards and potential of a fairly small-scale GIS directed to practical problem solving in a area of obvious policy relevance.

KINGSTON POLYTECHNIC
The data interrogation problem requires a considerable investment in research into suitable formats for data exchange and into data structures for spatial data storage and manipulation. At present research work on hierarchical spatial-data structures (eg. *quad-trees*) seems to be offering a suitable approach which should permit flexibility in the integration of multiple datasets at different scales/resolutions. Some work is also required on functional computer languages specifically designed for efficiently processing hierarchically

organised information of varying type. Some recent work on candidate languages for fifth generation computing may bear fruit in this area. Besides the advances being made on data structures and languages it is apparent that there is a major requirement for *artificial intelligence,* particularly knowledge-based systems, in the management of an access to spatial databases. Although work on knowledge-based systems is expanding rapidly at present primarily through the Alvey advanced information technology programme, it is not clear that there is likely to be any immediate 'technology transfer' to the geographic community.

BRIGHTON POLYTECHNIC, COUNTRYSIDE RESEARCH UNIT

The principal function of a geographic information system is practical. It is a tool for use in the collation and manipulation of basic spatial information for use as an aid to the decision-policy-making processes. In our experience it is the users' demands and objectives which are driving forward the research and development frontiers in this academic institution. Users' problems need immediate attention but in seeking solutions, the research frontier can then be encountered. We would suggest, therefore, that the identification and execution of research and development is a joint function between users and experienced research workers.

US DEPARTMENT OF THE INTERIOR, GEOLOGICAL SURVEY

The following have been identified by the Interior Digital Cartography Coordinating Committee (IDCCC) as technological trends or needed capabilities that will necessitate continued research and development of GIS technology for implementation in DOI programs:

1 The development of new and advanced *data structures* is needed to facilitate horizontal and vertical integration of data sets, improve storage and processing of text and attribute data, and allow data exchanges and communication between GIS systems.

2 The development of standards and procedures to facilitate *data exchange.* The various programs often have a need for data prepared for another purpose and the ability for quick and effective exchange would substantially reduce costs.

3 More effective *data base management systems* and advanced data storage technology, especially those designed for use with spatial data, are essential for the growth of this technology because of the size of many of the data bases involved.

4 The initial construction of data bases tends to be the most expensive aspect of GIS and will require continued research and development into advanced *data collection techniques.*

5 Advanced methods of *data communication* will be an important research topic to support field level applications of GIS.

6 A strong research effort is essential to develop *improved algorithms and techniques* for the handling of spatial data. In order to address the more complex and sophisticated analyses needed for most effective decision-making, new and enhanced techniques and capabilities to manipulate data will be required.

7 In a similar manner, *software transportability* would permit the exchange of software for specialised GIS functions or analytical modeling and would reduce development costs, enhance the applications of GIS technology, and increase the efficiency of data exchange.

8 *Microcomputer-based GIS* will be important to support the effective implementation of this technology in field locations. The rapid growth of microcomputers provides an opportunity for the transfer of this technology to the field level working environment.

9 Because of the demonstrated potential of remotely-sensed data, continued research and development is necessary on the *use of remotely sensed data with GIS technology.*

10 As GIS technology becomes less expensive and data bases more available, there will be many new applications within DOI, and it will be necessary to support continued development on *GIS applications* to these new areas.

The 10 areas discussed above could provide the focus for a generic GIS research and development program for the U.K. The users of GIS technology will be instrumental in identifying and defining specific R&D tasks needed for improving the application of GIS technology.

Key to accomplishing this research is the selection of a lead agency for research and development. Without a lead agency to coordinate and foster research some areas of investigation will not be covered, while others will be covered several times. Once a lead agency is selected, a budget for research and development must be secured. Then a determination must be made on the relative roles of government, industry, and academia in carrying out the research. Just as GIS applications are multidisciplinary, it is useful to involve researchers from a wide range of disciplines. The synergistic effects of having a variety of disciplines involved seems quite effective in solving GIS problems. Likewise direct interaction of the researcher with the users of GIS technology leads to results that can be incorporated into operational GIS programs.

H   Costs & Benefits

DEPARTMENT OF THE ENVIRONMENT

The Department recognises that there could be considerable benefit from the ability to link data from different systems both in terms of increased efficiency in the the collation and storage of information and to better inform policy issues. Information technology is beginnning to make a significant contribution in

(a) reducing the costs of handling information;

(b) increasing flexibility in analysing information; and

(c) improving the visual and cartographic representation of information.

GRAMPIAN REGIONAL COUNCIL

The Council is keen to take advantage of opportunities offered by technological change, but Government for its part must recognise the resources implications (capital and revenue) of introducing and extending new technology for the handling of geographic and other information bases. Priorities and a realistic programme for action must be decided, responsibilities within a programme allocated, and adequate funds made available.

Second, costs also form a constraint limiting developments in the handling of geographic information; local government expenditure is severely constrained and tightly controlled by central government. Developments in the handling of geographic information may require substantial investment over a long period before any great benefits ensure. Specific and substantial costs will also arise during the necessary period of overlap before old manual systems can be fully replaced by new automated systems. Expenditure more immediately related to improving services in the short term will inevitably be easier to justify in such circumstances.

CITY OF ABERDEEN

The establishment of a comprehensive system of compatible geographic information would have a clear commercial viability. Therefore the cost of establishing and operating such a system could, to some extent, be off-set by making the service available to a wide range of private and public sector agencies.

MERSEYSIDE COUNTY COUNCIL

The lack of corporate commitment to the project is caused mainly by the serious cut backs in revenue and resources, line managers naturally wish to protect their line mangement systems even if this is at the expense of support for corporate, strategic systems. This has proved especially true in local government where departments are organised as quasi-autonomous directorates. In such times of recession, when monitoring of performance becomes vital, the corporate management system is the first to be dropped because it has no direct responsibility for services and is not enforced by legislation.

SOUTH OXFORDSHIRE DISTRICT COUNCIL

If we can assume that there will be a dramatic increase in awareness of the benefits of computer map-based land and property information systems, that computers and computer systems will interconnect and that national agreement can be reached quickly as to the best sources for particular spatial data, THEN I am confident that local government will provide the lead that is so very necessary. It is possible in most local authorities for the mapping projects to be self-financing or cost justified with regard to benefits or increased efficiency.

MANPOWER SERVICES COMMISSION

A balance needs to be struck between the need to know and the cost of finding out. This is especially evident in relation to the MSC's perceived need for improving its local labour market information. Expenditure on programmes is enormous and there is an obvious need to ensure that it is well directed. Information already available needs to be supplemented, especially at the local area level, but ways in which this can be achieved through better management of the relatively low cost information which exists or could be obtained as a spin-off of administrative procedures needs to be pursued. This is being addressed by increased and improved use of distributed microcomputers to store and retrieve much of the information gathered through contacts with employers and local organisations. Such information, which will always be incomplete, is useful when set alongside the kind of hard statistical data held on NOMIS.

ORDNANCE SURVEY

GIS creation and development will be very expensive and time consuming in the short term and unless large quantities of data can be handled efficiently, there will be little or no return on the investment. Just trying to emulate the status quo by computer is an improper aim for GIS; it should be to make an increased amount of useful information available to authorised users more quickly and more cheaply than is available by manual methods today, and in ways and combinations that have hitherto been considered impractical.

NATURAL ENVIRONMENT RESEARCH COUNCIL

It is generally true that, the more widely data are to be made available, the greater the cost both in acquisition and data management. If data is to be made available for a range of purposes, shortcomings in accuracy, consistency and timeliness can quickly become intolerable. Even if data are kept accurate, consistent, and up-to-date, datasets vary greatly in their objectivity. Some are unambiguous, and require little interpretation — eg a digitised coast line or bibliographic index. Others many require expert scientific interpetation, and might even be misleading if accessed by someone with insufficient background. It is probably true to say that NERC's data is predominantly in this second category. Such difficulties in interpretation by third parties should never be used as an excuse for not trying to collect data in as widely applicable a manner as possible, so that it can then be disseminated. As an example BGS has had some success in international co-operation with other bodies on standardisation of geological terms.

Within NERC itself, many datasets are managed locally by the scientists who collect them, and make them available to other scientists on request. Other datasets are managed centrally for NERC as a whole, with some of these providing an external service. Bodies external to NERC invariably obtain such data via a human interface and do not log on to NERC's computers without the intervention of NERC staff. These matters are relevant to where data are managed. Whether data should be physically held locally, centrally to NERC, or centrally external to NERC is a pure matter of the trade-off between technological practicability and convenience. It wastes storarge to duplicate data at different locations, and can cause problems in the duplicates becoming out of date. Against this must be reflected the current limitations of data communication when transferring substantial quantities of data. NERC is a dispersed organisation and is heavily reliant on computer networking, and NERC and university scientists freely use both local and remote computers which are interconnected by the Joint Academic Network (JANET). In many cases scientists will use JANET to obtain subsets of data which are then worked on intensively using local computing facilities.

There would be great advantage to NERC in being able to obtain data — eg digital land outlines — that are held by external bodies. In a number of cases BGS data sets are derived from databases which are external to the UK and controlled by international agreements (eg World Geomagnetic Data Centres). In such cases local databases must conform to international agreements and formats. Data exchange and sharing is also dependent on the availablity of data indexes and agreements on standard formats. The technical problems associated with data exchange and communications can largely be overcome using available technology. The challenges to be met are primarily of an organisational character and concerned with such questions as agreements on standards, agreements on responsbilities, fair pricing structures and copyright.

HALIFAX BUILDING SOCIETY

We foresee considerable benefit from being able to access, normally on-line, a range of spatial information. Obvious areas of interest are access to the Land Registry and digital maps, whilst other information such as census data and planning applications would also be useful.

The main constraints would be the validity of the data, the reliability of the access procedures and the cost of access.

INTERNATIONAL COMPUTERS LTD

By far the largest cost for the typical user is in the data collection, validation and organisation of the underlying spatial data or geographic backcloth against which the data available to the user is to be analysed and displayed. The cost of data capture has been (and remains) the most effective barrier to most organisations. Traditional hand digitising in a labour intensive operation and until recently alternatives were unsatisfactory in quality. Recent advances in scanning systems have allowed the economic capture of quality images, but these have limitations in use within systems.

HEALEY & BAKER LTD

We recognise, however, that in the past the collection of such information may not have been effective in a cost sense, but it seems to us that there is now a very much wider market for research data than formerly, and that those who make use of such information would be willing to make a substantially greater contribution to the costs of collection. Whilst we cannot say that the charges that could be made for such data would fully cover the cost of collection, it seems to us that users would not be deterred by substantially higher charges than were previously the case and that, in so far as costs were not fully met, the shortfall should be viewed as a necessary contribution to overall knowledge as to trends in commercial activity throughout the United Kingdom.

SOUTH WESTERN ELECTRICITY BOARD

There is no doubt, however, that the Government would see the reward for such support. The oportunities provided to improve the efficiency of many aspects of UK business management systems is self-evident. It is also important however to consider the export potential of development work in this field. During the development of procedures in SWEB many foreign visitors from a variety of Utilities have visited the Board to evaluate the work being undertaken.

THAMES WATER

It is possible in theory that an electronically held record of geographic information can exist in one place only and be accessed nationwide as required. The great advantage is that revision need only be made in one place in order to keep the record permanently up-to-date. It is pleasant to contemplate future on-line access to any OS map, knowing it will be completely up-to-date, and to be able to superimpose underground utilities, land ownership boundaries, and planning data, for example. There are many organisations which now have immediate on-line access to centrally held data, the banks and some building societies being among the better known examples. However, the cost of such a utopian system might well be prohibitive, Futhermore, it seems likely that paper copies will alway be required for wall diagrams, legal records, meetings, site visits, etc. Nevertheless, it is desirable that the likely costs of on-line access to centrally held records, based on a nationwide system, should be established soon, before a significant number of major organisations become committed to alternative development programmes; until such costs are known it seems likely that Thames Water will continue with a local development programme in conjunction with other area utilities and Local Authorities.

LOCAL AUTHORITIES ORDNANCE SURVEY COMMITTEE

Local authorities themselves are under severe economic restraints, and are therefore unable to progress as far and as fast as they would wish in certain areas, because of cash limits. It is inevitable that in such a climate the use and type of geographic data collected by the authority is not seen as such a high priority as other local authority services. Therefore the increased use by local authorities of geographic data, especially digital map data is likely to be conditional on the costs involved, and the cash limits placed on local authorities.

Therefore, local authorities request that the Government take into account the resources implications (capital and revenue) or introducing and extending new technology for the handling of geographic and other information bases.

ROYAL SOCIETY OF EDINBURGH

The cost of acquiring data is a major concern. It appears increasingly to be Government policy to seek to recover from users as much as possible of the cost of producing. For academic researchers, such charges can be met only from limited research funds, which are themselves under increasing pressure as a result of the reductions in expenditure on higher education. In the past, such information was often available in some detail at no or little cost; this enable researchers to undertake preliminary studies to establish the quality of the data and their appropriateness. In many instances, this approach is no longer possible and the need to commit limited funds in advance may subseqently lead to abortive inquiries. Alternatively, it may lead to no research being undertaken or, as in socialist countries where (for different reasons) geographic data are not available, to the replacement of useful research about the real world by purely theoretical studies.

UNIVERSITY OF NEWCASTLE, CENTRE FOR URBAN AND REGIONAL DEVELOPMENT STUDIES

The cost problem of data access is another which has tended to increase directly as a result of the smaller areas for which data is coming available (as costing is often per area). Departmental policy too often stresses a more 'commercial' approach − yet this in interpreted solely as the basis for a price increase and not for developing new 'products' (such as city classification) or even adopting a price-cutting strategy to encourage custom (the US Census data is far cheaper and more widely purchased than its British equivalent). Again, if the Domesday project develops in the way suggested then the need for more competitive pricing may become more evident.

LOWELL STARR, US DEPARTMENT OF THE INTERIOR, GEOLOGICAL SURVEY

Are digital cartographic GIS's cost effective? From the experience of USGS the answer is an unqualified, yes. Although the start-up, front-end, initial costs are very high in terms of all resources, the benefits are enormous in terms of the ability to analyze, maintain, and update spatial data. Once a user has developed a custom product and seen the capability to do specific analysis that *answers* the questions, that user will never again be satisfied with anything less. The cost to update, maintain, and output data becomes minimal − which in most cases is just the opposite of standard data base application.

I Role of Government

MINISTRY OF DEFENCE, MILITARY SURVEY

The national initiative discussed in answer to qustion 8 requires the full authority of Central Government. It must address the identification of requirements and definition of standards for exchange and must direct future developments. It will need to create the infrastructure for data exchange. None of this can be achieved in a realistic time frame without the backing of Central Government in both financial and legislative terms. Failure to take such an initiative now will result in continued unco-ordinated development, lack of interoperability, duplication of effort and the creation of entrenched positions to the extent that future control will prove difficult or even impossible.

INLAND REVENUE (VALUATION OFFICE)

There is an obvious need for national standards for location referencing and data exchange. In our opinion Central Government should assume responsibility for co-ordination and for the achievement of national, and perhaps international standards.

We have suggested the establishment of a national register of property information by the linking of existing databases. This would enable Government to set an example.

NATIONAL JOINT UTILITIES GROUP

The major obstacle to the change from conventional mapping to a digital base is the cost involved. However, it is the key to most of the information systems used in government (national and local), land registry, utilities, transport etc. The ultimate cost benefits of the systems can already be seen but this is already being eroded by a growing duplication of the initial work involved. Central Government should prevent this erosion by funding the rapid provision of background digital mapping of the country. Once this is done each user will fund their own transfer of records.

BRITISH GAS CORPORATION

Local authority boundaries generally do not coincide with the Utility operating boundaries, although the Utilities have recognised their own boundary problems and are developing a common approach, there may be problems linking with local authorities. The 450 local authorities in the UK have a high degree of autonomy subject to local political constraints. Consequently if there is a need to adopt common data exchange standards and protocols to achieve electronic exchanges of plant records or notification of intended works in streets, central government influence on local authorities may be essential. In the report from the Review of the PUSWA Act 1986, Professor Horne may well recommend the development of improved notification procedures based on electronic office technology. No doubt central government will recognise this particular aspect in their response to the Horne proposals. Also recommended by Horne (see recommendation) is that the government provide financial assistance to local authorities to create the new exchange system. At this early stage the utilities understand that little progress can be made by local authorities on this aspect without legislation and financial assistance.

ASSOCIATION OF DISTRICT COUNCILS

The primary role of central Government must be a commitment to a regular and systematic collection of information necessary for the development of local and national policies and plans. The census must be regarded as sacrosanct, and reintroduction of the mid-term census and the census of distribution, as indicated above, would be helpful.

Building upon the foundatation of this commitment, Government should identify the need for information, commission research into data gathering, storage, retrieval and distribution techniques, and take action in implementation of the research findings.

The Association has argued earlier in this memorandum a further role for Government in increased support for the programme of digitised mapping and software development to render that programme more useful and less expensive for local authorities. An unduly high price for digital mapping during the development period will depress the market and make future recovery of development costs more difficult.

MIDDLESBROUGH BOROUGH COUNCIL

To summarise — the government may have a useful role:

(i)   to research and devise workable standard spatial reference systems for general use nationally;

(ii) to act as a clearing house and monitoring service on the existence of information systems and experience gained from them;

(iii) to provide general training, updating and awareness facilities for those working in this field.

SOUTH OXFORDSHIRE DISTRICT COUNCIL

Nothing can be achieved nationally without firm Government direction, courage and the input of the very considerable resources necessary.

EXETER CITY COUNCIL

The linking of information systems will lead to a need for new standards for referencing of spatially based information, the diversity of requirements of users must be equated against corporate information needs, This implies a need for nationally adopted standards for location referencing to enable on-line access to property data, led by Central Government with the Local Government case being represented by LAMSAC. However, the difficulties of referencing information generally held on maps, eg underground utilities, in a manner suitable for computer systems will not be resolved solely by property referencing. To ensure compatibility and access to property based and 'linear' types of information a digital system is required. Again it would seem appropriate for Central Government to lead such development, perhaps through the Ordnance Survey, in this respect their programme of digitisation needs to be continued perhaps at a quicker rate, particularly in urban areas.

CHESHIRE COUNTY COUNCIL

The need for commonly agreed standards — perhaps a code of practice — has already been highlighted by the exchange of information within Cheshire between the County Council, District Councils, statutory undertakers and the Ordnance Survey. A Code of Practice should cover such matters as responsibility for digitising data: the scale at which this should take place and other technical aspects. Principles relating to charging, security and confidentiality would also need to be covered as well as an agreed geographical framework within which data would be stored.

Agreement between local users would be of limited value if there was still problems when using national data. Central Government and national agencies need to be responsive to local requirements. The Department of the Environment recently digitised local authority boundaries at a scale which was too small for local use. National data is often released in units which are 'local' from a national standpoint but are too coarse for local authority use.

These factors point to the desirability of Central Government taking an initiative (in consultation with a national body which could represent the local viewpoint — like LAMSAC) to develop national standards and a nationally agreed Code of Practice relating to geographic information. This initiative need not start from scratch but could take into account other evolving standards. If a generally accepted code can be defined, the exchange of spatial information should become easier and more commonplace.

INTERNATIONAL COMPUTERS LTD

Central Government departments are already major providers and users of geographic information, so it will be in the overall interest of Government that development is cost effective and meets its own substantial user needs. The creation costs of major GIS will be very high and Government will have to contribute to these costs either by a firm commitment to purchase commercial packages or by direct funding of development. Apart from Government the only other major users who are likely to have the financial resources to fully develop GIS are the public utilities, oil companies and to a lesser extent the local authorities. If the Committee of Enquiry recommends that GIS should be handled only on a local scale then the case for significant Government involvement is weakened. A central government commitment to make its own data available for incorporation into major GIS is a pre-requisite for the long term benefits of GIS to be realised.

BRITOIL

Central Government is the only body which can fill the co-ordinating and legislative role in terms of setting and ensuring the implementation of standards, specification etc. There are already a number of bodies which receive government funding such as Department of Energy, MIAS, Met Office, BGS, etc where a lead could be shown in terms of standardisation and co-ordination. Central Government also has to ensure that the national mapping agencies take note of the long term use of data. For example an oil company will establish the position of a platform or pipeline to 5 to 10 m but the Hydrographic Department only require the information for charting purposes to 25 to 200 m depending on chart scale.

GEOPLAN (UK)

The role of Central Government is vital to the establishment of standard geographical boundaries, and Geoplan welcomes any government move towards the use of post code based geography for the SAS, and indeed any other large scale government surveys. Increased efforts by government to co-ordinate with other bodies both private and public would also be welcomed by Geoplan.

GEC RESEARCH

Whilst the long term commercial potential of GIs is high the capital investment required now is prohibitive to commercial enterprises. We are dealing with the efficient exploitation of a national resource and a lead should therefore be given by central government. Indeed many Government departments and ministries will benefit from spatial information technology development, also the military potential is enormous.

HALIFAX BUILDING SOCIETY

The role of central government should be to encourage standards and competition, and thereby generate the fastest possible growth in the areas of information provision and networking. Subsidies and incentives could be given to 'early participants' in order to encourage organisations to become involved at any early stage and get projects off the ground more quickly.

COUNTY SURVEYORS SOCIETY

The role of central government should be to see that anybody appointed to co-ordinate developments in geographic information systems should have adequate funds and facilities to carry out the task with maximum speed and efficiency. The cost of setting up and running spatial referencing systems is likely to be small compared to the benefits arising from their use.

It is essential that a Standing Advisory Committee be established to identify users requirements and determine policy.

Central government sponsorship is essential to support and co-ordinate a national referencing system.

ROYAL GEOGRAPHICAL SOCIETY

The Government must take the lead through appropriate policies, by funding research, development and education and the encouragement of the skills required. There must be a Board, for which one Minister must be responsible, to co-ordinate and promote the development of geographical information systems nationally.

A geographic information system at national level will not happen without substantial Government commitment and involvement. An example of the importance of such commitment is the case history of the recommendations in the Report on General Information Systems for Planning. In 1972 the Government gave encouragement to local authorities to adopt this system for planning but then backed off probably due to expenditure implications. The result is the existence now of a large number of incompatible systems.

Government must lead as it is already the major provider of all the data with the national coverage that will be required in geographic information systems. This lead must be given not only by co-ordination and organisation but also by defining appropriate policies and providing adequate funding. It is most important that co-ordination and development in this field be the responsibility of a single Government minister.

HOUSE BUILDERS FEDERATION

HBF as a user of certain specific categories of geographical information primarily for the purpose of making effective representations, comments and submissions on the development plan system, identifies a clear role for central Government in the development of improved geographic information systems. The Federation considers that the overall planning and provision of services and land use policies could be greatly improved by the establishment of a central co-ordinating agency within Government, which would effectively pull together the wide range of disparate geographical information that is now available from both public and private sector sources.

STANDING COMMITTEE OF PROFESSIONAL MAP USERS

We consider that the survey and production of maps is best carried out by the Ordnance Survey. Government or Government-related organisations already have the majority of additional information which can usefully be shown on maps — we refered to such things as pipelines and services, grades of agricultural land, etc. We suggest the problem is one of co-ordination which should be studied by undertaking a small number of pilot schemes in various parts of the country to be monitored and evaluated by a small number of interested users, OS and Government personnel.

ROYAL SOCIETY OF EDINBURGH

The Society believes that the Government has a key role to play in the adoption of modern methods of handling geographically-referenced data. Only the Government can fund the collection of data in appropriate forms and the necessary basic research. The role of government is particularly important in the development of geographical information systems. The Society greatly regrets the termination of the Rural Land Use Information System (RLUIS) project, funded by the Standing

Committee on Rural Land Use. This unique collection of data for the Dunfermline and Kirkcaldy Districts in Fife was supplied both by departments and agencies of central government and by the local authorities, and had great potential for the examination of problems of handling a wide variety of data and of developing and adopting geographical information systems.

ROYAL TOWN PLANNING INSTITUTE

This requires that central government takes a positive initiative in encouraging the development and adoption of a standard format. This is not a matter where vague 'market forces' can be left to control development. It is too important for that. Central government could also encourage the rapid adoption of the chosen format by both requiring the data it collects to be presented using that format and by itself using that format for the data it uses. It is a prerequisite of a geographic system utilising new technology that there be a digitised map base to which other data can be attached and perhaps the key role of central government is thus in the funding of the Ordnance Survey and its operations to ensure adequate and speedy provision of the digitised 'geographic base'. Without in any way prejudicing the requirement that the Ordnance Survey be as profitable as possible, the injection of funds to ensure geographic cover in digitised map and plan form at 1/1250, 1/2500, 1/10,000 and 1/50,000 scales is essential if this task is to be undertaken within a realistic time scale. At current levels of funding and technology it is understood that the O.S 'best estimate' is that it will take some 30 years to complete a programme of digitising the 1/1250 and 1/2500 scale plan cover.

UNIVERSITY OF BRISTOL, SCHOOL FOR ADVANCED URBAN STUDIES

We believe that there are strong economic as well as 'public interest' arguments to justify central government taking a greater role in this areas. At present there is much wasteful duplication of effort — particularly amongst secondary users of major sources — in both the public and private agencies attempting to utilise data sources to support their functions. A great deal of research time is spent in local authorities, government departments, private firms and academic institutions trying to reconcile conflicting spatial frameworks, as well as trying to find out what data are available on what basis. Much of this effort takes place in isolation so that little collective benefits is gained from individual investment or expertise. On this basis we consider that there are longer term cost savings arguments for the central coordination of research and development in the geographic information systems arena. We believe that data collected by any public agency can and should be regarded as a public good and therefore publicity available insofar as is compatible with the protection of confidentiality. The cost justification to a specific agency of organising and referencing data to facilitate spatial analysis should be seen not purely in terms of the direct utility of that data to the collecting agency, but within the broader frame of its collective utility as a public resource.

UNITED STATES, DEPARTMENT OF THE INTERIOR, GEOLOGICAL SURVEY

Discussion on the role of government in developing geographic information systems often hinges on the desirability and utility of public domain software. In the United States, software developed by the government is available to the public at a modest cost. Users who desire to obtain a GIS at a low unit cost often argue that is is the proper function of government to develop such systems particularly if the software is in direct support of a government mission and other uses are considered as serendipity.

The other contention is that the private sector can do a better job of developing new systems and government should procure and use the best available system. If one considers the cost to the government for original development and subsequent maintenance, it is argued that software procured from the private sector has a comparable unit cost to public domain services.

The Geological Survey sees a role for both government developed and private sector geographic information systems. In many cases government scientists have the best understanding of particular problems that need to be solved and are best equipped to develop the solution. Although the solution may first attempt to utilize proprietary software, if that software is not adequate new capabilities may need to be developed by the government. the various situations need to be evaluated on their own merits; blanket statements that government should not develop GIS software cannot be supported.

J Co-ordination

ORDNANCE SURVEY

The initial development of GIS occurred through the enthusiasm of individuals and even now awareness of GIS potential is confined to relatively few organisations and people. There is the danger that this can lead to different and perhaps incompatible developments. As more people become aware of the benefits of GIS, more development is likely to occur and it is at this early stage in the development process that new developers need advice and also that common standards should be introduced. If further development is allowed to take place completely free of common standards, then it will be more difficult to direct the total effort towards a common cost-effective goal.

If GIS are to be successful, they must relate several datasets together through some common reference system. Many of the datasets may not have been related in any obvious way in the past and they will cover a wide spectrum of interest and disciplines. Ordnance survey believes that the common reference system should be the National Grid. However, many of the GIS designers involved may not be familiar with this reference system and, in the absence of advice, may choose entirely different approaches. This will make inter-relationship between datasets more difficult.

For these reasons, Ordnance Survey argues that there should be a co-ordinating body to whom developers of GIS can turn for advice and information. To avoid unnecessary start-up costs, Ordnance Survey considers this co-ordinating role should be taken by an existing organisation. As Government departments will be major contributors to and users of GIS, it seems appropriate that a Government body should provide the co-ordinating service. Ordnance Survey sees the major functions of the co-ordinating body as being:

- the provision of a forum for the discussion and exchange of development information;

- the provision of advice to creators and users of GIS in both the public and private sectors;

- the formulation of data exchange standards; and

- the maintenance of a register of important geographic information systems.

Commercial considerations obviously dictate that there will be competition to develop software but there is a role for a co-ordinating body to ensure that money is not wasted on unnecessary research and that basic development is not being repeatedly duplicated.

NATIONAL REMOTE SENSING CENTRE

To support and maintain research efforts we propose that the Alvey Directorate should be taken as a guide, and that a coordinating committee be established to oversee and control research and development efforts so as to achieve maximum benefits by avoiding duplication and waste. The coordinating committee should draw its membership from all sectors of the R & D community and have as its ultimate aim the production of a UK geographical information system to serve the needs of UK users. Such a system would also need to meet the requirements of overseas users.

The Information Handling Working Group would be willing to provide assistance in the work of coordination.

If such a committee were to be established it should endeavour to coordinate the activities of all groups, large and small, working in the fields of interest that we have already outlined. We believe that there is considerable expertise scattered throughout the UK in small groups which may be overlooked if attention is focussed too closely upon the development of large-scale systems in one or two large centres. The availability of spatial datasets via the proposed demonstration GIS project might be used to draw in this scattered expertise in a meaningful way.

ROYAL INSTITUTION OF CHARTERED SURVEYORS

Throughout our evidence we have stressed the importance of standards and co-ordination and we believe there is a clear need for a central national body to achieve this. This central body should also aim to set national objectives for the development and use of spatial data systems.

Whilst we have no firm views on what form this central national body should take, we believe strongly that it should be independent and should have full Government support, with the Ordnance Survey taking a leading, although not dominant, role and with representatives drawn from experienced user groups.

It is important that this body is not simply a 'talking-shop' and that its members should be of sufficient seniority to be able to commit their organisations to specific technical decisions and courses of action, even if not to spending money.

It is equally important that members should be largely drawn from organisations, whether public, commercial or academic, which have direct practical experience of spatial information systems and are aware of the problems and opportunities. They need to be cartographically and spatially literate, with the ability to anticipate the problems and foresee the consequences of merging data sets of widely differing qualities and standards.

In addition to the central national body there will be a need for regional and/or specific-interest user groups and working parties. These would be charged with developing and agreeing national standards along the lines indicated earlier in our evidence. They could also examine specific development areas, recommending development projects where this seemed desirable.

The central body might also give some consideration to the possibility of a national cost-benefit analysis for the use of digital geographic information systems, linking this to the declared national objectives.

NATURAL ENVIRONMENT RESEARCH COUNCIL

In the absence of any central direction developments in geographic data handling are likely to be characterised by a lack of coherency. A national development strategy is needed that allocates responsibilities and enables progress to be made in such critical areas as standards, data exchange mechanisms and copyright. It is recommended that a National Programme Board for Handling Geographic Information is established and chaired by DOE. Co-ordinating agencies should be nominated to work with the Board in key sectors (eg small scale and large scale applications, technology). An early task for the Board would be to prepare a national plan for the development and application of geographic handling systems.

SOUTH WESTERN ELECTRICITY BOARD

There is no doubt that the need to co-ordinate development in the field of digital map records is very important. The Board consider that there are a number of different levels on which this should operate. At a national level there appears to be a need for 4 types of development agencies:

(a) to promote the development of digitally based background maps, co-ordinating the role of Ordnance Survey and Mapping Contractors;

(b) the continuation of the NJUG approach to co-ordinate the inter-utility developments, but more closely aligned with centrally expressed views of Local Authorities and the Departments of Transport and the Environment;

(c) the independent co-ordination of development of the Utilities undere the wing of, for example, the Electricity Council for the Electricity Supply Industry and similarly for Local Authorities;

(d) the co-ordination of technological development of CAD system development, automatic scanning systems and information processing and transfer.

At a local level it is the Board's view that considerable benefits arise from co-ordinated development between Utilities and Local Authorities. From such can come specific joint location developments such as has been achieved in Dudley and at Taunton.

LAMSAC

Co-ordination and liaison on spatial locational referencing was at one time carried out by the CCLR (Co-ordinating Committee on Locational Referencing). This was run by the Department of the Environment. It was the ideal forum for all matters relating to spatial referencing and in our opinion is sadly missed. It was through the Committee and the LOGIS (Local Government Information Systems) research funding that the Ordnance Survey Restructuring project was initiated.

Advantage would accrue if the Committee was reinstated with updated terms of reference.

At the supposed Centre of Excellence, the Ordnance Survey, in conjunction with the central government co-ordinating body (CCLR) should be responsible for guiding if not initiating research and acting as advisors to the relevant funding bodies who support digital mapping data research.

SOUTH TYNESIDE METROPOLITAN BOROUGH COUNCIL

Insofar as organisation arrangements are concerned, the critical factors will be: the level and source of resources made available; the scale of operation; and, the nature of promotional/control agencies. The Committee will appreciate that the level of resource commitment − particularly to development and installation − for a national network system will be high. Moreover, to meet the diversity of requirements it will be necessary to generate and coo-ordinate responses at national level, either by Government or through a Government-appointed agency. A first stage would involve a feasibility study into the technical and cost implications of provision of a national data transmission system which satisfactorily meets the requirements of potential users (assuming that all such users are identifiable).

BRITOIL

The solutions to these problems are very difficult because each data type has its own characteristics, the user or originator has his own specification, and the number of spatial relationships can be very large. A great deal of work will be required to try and achieve standardisation of formats, accuracy criteria, etc. Bodies such as United Kingdom Offshore Operators Association have tackled this for very specific purposes such as exchanging positional data and could highlight some of the problems encountered.

In order to achieve this, Central Government has to take the lead by defining and co-ordinating any perceived National requirement for such databases. The problems of access to, availability of, confidentiality of, accuracy of, and duplication of data have to be addressed at an early stage. This will probably require the establishing of standardised procedures by legislation.

INTERNATIONAL COMPUTERS LTD

How to ensure that UK skills and solutions are fostered to provide internationally competitive products:

This can only be done through support.

- Research and Development funding.
- Support for innovative projects.
- Co-ordination of projects, to bring together teams of imaginative user organisations and supplier organisations.
- Using public sector procurement to "force" the development of UK industry.
- Playing an active role in international standards formulation.

How to ensure planning is effective in a field of rapidly changing technology.

Here again, co-ordination is the key.

- Taking a broad view from fundamental R&D through to product development.
- Encouraging good system design practices and the promotion of standards.

- Ensure that geographic data considerations are included in related research (eg Alvey programme).

HOUSE BUILDERS FEDERATION

There would seem to be considerable scope for Government departments, such as the Department of the Environment and the OPCS, to liaise with many of these private sector organisations in order to produce mutually compatible data sets which can be developed to generally raise the overall standard of data availability in fields such as housing and employment. It may well be that some co-ordinating agency is required to undertake this function. There is clearly a great deal of information available to both public and private sectors and the quality of forward planning and service provision would be greatly enhanced by some sort of central agency that seeks to draw together much of this information in a compatible and readily usable manner.

ROYAL SOCIETY

The lack of coordination in the collection and archiving of spatial data is undoubtedly leading to unnecessary duplication of effort and expenditure, proliferation of different standards, uncertainty about copyright and charging policies, and a general ignorance by groups and organisations of the extent of work being carried out elsewhere. A start has been made towards remedying some of these undesirable situation situations by, among other things, establishing the Committee of Enquiry; but an appropriate trans-departmental administrative structure headed by a Minister and with its own budget and secretariat is neded to coordinate and direct effort. This structure would include the government departments that collect data, such as the Ministry of Defence and the OS, and would also involve the research councils. It would not only address the problems mentioned but would also determine research priorities and channel funds in the right direction.

ROYAL GEOGRAPHICAL SOCIETY

There should be a *coordinating Board* at the highest level with wide executive representation from Government ministries, Research Councils, the Universities, national Utilities and Local Authorities and Industry. On this Board, and its committees reflecting particular areas of application, it is most important that users other than those in the data collecting departments are included. Only in this way will narrow departmental interest and unwillingness to cross departmental and organisation boundaries be overcome.

There should be a widely distributed newsletter to all interested organisations and agencies, such as that produced by the United States Geological Survey, to show progress on common geographic information standards, methods of data aquisition and the use of geographic information data.

There should be some central contact point for expert advice for potential users of such information.

Printed in the United Kingdom for Her Majesty's Stationery Office
Dd 290558 C15 6/88 3936 12521